# THE
# ILLUSTRATED
# ART OF WAR

SUN TZU

# THE

# ILLUSTRATED

# ART OF WAR

THE DEFINITIVE
ENGLISH TRANSLATION BY
SAMUEL B. GRIFFITH

OXFORD

Oxford University Press

Oxford University Press, Inc., publishes works that further Oxford University's
objective of excellence in research, scholarship, and education.

Oxford   New York
Auckland   Cape Town   Dar es Salaam   Hong Kong   Karachi
Kuala Lumpur   Madrid   Melbourne   Mexico City   Nairobi
New Delhi   Shanghai   Taipei   Toronto

With offices in
Argentina   Austria   Brazil   Chile   Czech Republic   France   Greece   Guatemala
Hungary   Italy   Japan   Poland   Portugal   Singapore   South Korea
Switzerland   Thailand   Turkey   Ukraine   Vietnam

This edition published in the United States of America by
Oxford University Press, Inc., 2005
198 Madison Avenue, New York, New York 10016
www.oup.com

Oxford is a registered trademark of Oxford University Press

Library of Congress Cataloging-in-Publication Data

Sunzi, 6th cent. B.C.
  [Sunzi bing fa. English]
  The illustrated Art of war / Sun Tzu ; the definitive English
translation by Samuel B. Griffith.
    p. cm.
  ISBN: 978-0-19-518999-5 (hardcover : alk. paper)
  I. Griffith, Samuel B. II. Title.
  U101.S9513 2005
  355.02--dc22
                                            2005010651

Conceived, created and designed by
Duncan Baird Publishers, London, England
Managing Editor: Chris Westhorp
Managing Designer: Dan Sturges
Commissioned Calligraphy: Yukki Yaura

Typeset in Garamond
Colour reproduction by Colourscan, Singapore
Printed in China by Regent

10 9 8 7 6 5 4

**Note:** The text in this edition is reproduced from the classic 1963 edition. Additional material
relates to the illustrations. However, only two out of the four appendices are reproduced from
the 1963 edition, which is available from Oxford University Press ISBN 978-0-19-518999-5.

# CONTENTS

# FOREWORD

Sun Tzu's essays on 'The Art of War' form the earliest of known treatises on the subject, but have never been surpassed in comprehensiveness and depth of understanding. They might well be termed the concentrated essence of wisdom on the conduct of war. Among all the military thinkers of the past, only Clausewitz is comparable, and even he is more 'dated' than Sun Tzu, and in part antiquated, although he was writing more than two thousand years later. Sun Tzu has clearer vision, more profound insight, and eternal freshness.

Civilization might have been spared much of the damage suffered in the world wars of this century if the influence of Clausewitz's monumental tomes *On War*, which moulded European military thought in the era preceding the First World War, had been blended with and balanced by a knowledge of Sun Tzu's exposition on 'The Art of War'. Sun Tzu's realism and moderation form a contrast to Clausewitz's tendency to emphasize the logical ideal and 'the absolute', which his disciples caught on to in developing the theory and practice of 'total war' beyond all bounds of sense. That fatal development was fostered by Clausewitz's dictum that: 'To introduce into the philosophy of war a principle of moderation would be an absurdity—war is an act of violence pushed to its utmost bounds.' Yet subsequently he qualified this assertion by the admission that 'the political object, as the original motive of the war, should be the standard for determining both the aim of the military force and also the amount of effort to be made'. Moreover, his eventual conclusion was that to pursue the logical extreme entailed that 'the means would lose all relation to the end'.

The ill-effects of Clausewitz's teaching arose largely from his disciples' too shallow and too extreme interpretation of it, overlooking his qualifying clauses, but he lent himself to such misinterpretation by expounding his theory in a way too abstract and involved for concrete-minded soldiers to follow the course of his argument, which often turned back from the direction which it seemed to be taking. Impressed but bemused, they clutched at his vivid leading phrases and missed the underlying trend of his thought—which did not differ so much from Sun Tzu's conclusions as it appeared to do on the surface.

The clarity of Sun Tzu's thought could have corrected the obscurity of Clausewitz's. Unfortunately, Sun Tzu was only introduced to the West, by a French missionary's summary translation, shortly before the French Revolution, and although it appealed to the rational trend of eighteenth-century thinking about war its promise of influence was swamped by the emotional surge of the Revolution and the subsequent intoxicating effect of Napoleonic victories over conventional opponents and their too formalized tactics. Clausewitz began his thinking under the influence of that intoxication, and died before he could complete the revision of his work, so that this lay open to the 'endless misconceptions' he had foreseen in his testamentary note. By the time later translations of Sun Tzu were produced in the West, the military world was under the sway of the Clausewitz extremists, and the voice of the Chinese sage had little echo. No soldiers or statesmen heeded his warning: 'there has never been a protracted war from which a country has benefited'.

There has long been need of a fresh and fuller translation of Sun Tzu, more adequately interpreting his thought. That need has increased with the development of nuclear weapons, potentially suicidal and genocidal. It becomes all the more important in view of the re-emergence of China, under Mao tse-tung, as a great military power. So it is good that the task should have been undertaken, and the need met, by such an able student of war and the Chinese language and thought as General Sam Griffith.

My own interest in Sun Tzu was aroused by a letter I received in the spring of 1927 from Sir John Duncan, who was commanding the Defence Force which the War Office had dispatched to Shanghai in the emergency arising from the advance of the Cantonese armies under Chiang Kai-shek against the Northern war lords.

Duncan's letter began:

I have just been reading a fascinating book 'The Art of War' written in China 500 BC. There is one idea which recalled to me your expanding torrent theory: 'An army may be likened to water: water leaves dry the high places and seeks the hollows; an army turns from strength and attacks emptiness. The flow of water is regulated by the shape of the ground; victory is gained by acting in accordance with the state of the enemy.' Another principle contained in the book is acted upon by Chinese generals of today; it is 'the supreme art of war is to subdue the enemy without fighting'.

On reading the book I found many other points that coincided with my own lines of thought, especially his constant emphasis on doing the unexpected and pursuing the indirect approach. It helped me to realize the agelessness of the more fundamental military ideas, even of a tactical nature.

Some fifteen years later, in the middle of the Second World War, I had several visits from the Chinese Military Attaché, a pupil of Chiang Kai-shek. He told me that my books and General Fuller's were principal textbooks in the Chinese military academies—whereupon I asked: 'What about Sun Tzu?' He replied that while Sun Tzu's book was venerated as a classic, it was considered out of date by most of the younger officers, and thus hardly worth study in the era of mechanized weapons. At this, I remarked that it was time they went back to Sun Tzu, since in that one short book was embodied almost as much about the fundamentals of strategy and tactics as I had covered in more than twenty books. In brief, Sun Tzu was the best short introduction to the study of warfare, and no less valuable for constant reference in extending study of the subject.

B.H. LIDDELL HART

# PREFACE

S su-ma Ch'ien, whose monumental *Shih Chi* (*Historical Records* or *Records of the Historian*) was completed shortly after 100 BC, tells us that Sun Wu was a native of Ch'i State who presented his 'Art of War' to Ho-lü, king of semi-barbarous Wu, in the closing years of the sixth century BC. But for hundreds of years Chinese scholars have questioned the reliability of this biography; most agree that the book could not possibly have been written when Ssu-ma Ch'ien said it was. My study of the text supports this opinion and indicates a date of composition during the fourth century BC.

Sun Tzu's series of essays does not merit our attentive interest simply as an antique curiosity. 'The Art of War' is much more than that. It is a thoughtful and comprehensive work, distinguished by qualities of perception and imagination which have for centuries assured it a pre-eminent position in the canon of Chinese military literature.

This first of the 'martial classics' has received the devoted attention of several hundred Chinese and Japanese soldiers and scholars. Among the most distinguished was Ts'ao Ts'ao (AD 155–220), the great general of the Three Kingdoms period and founder of the Wei dynasty. During the eleventh century his comments on the text together with the remarks of ten respected T'ang and Sung commentators were collated in an 'official' edition. In the last quarter of the eighteenth century this was revised and annotated by Sun Hsing-yen, a versatile scholar and celebrated textual critic. His edition has since been considered standard in China and my translation is based on it.

Sun Tzu was first brought to the attention of the Western world by a Jesuit missionary to Peking, Father J.J.M. Amiot, whose interpretation of 'The Art of War' was published in Paris in 1772, towards the close of a period during which the imagination of French artists, intellectuals, and craftsmen had been significantly influenced by the newly discovered and exciting world of Chinese arts and letters. Contemporary journals published favourable reviews and Amiot's work was widely circulated. It was again published in an anthology in 1782. Possibly this was read by Napoleon, as one Chinese editor has recently affirmed. As a young officer the future emperor was an avid reader; it is unlikely that these unique essays would have escaped his attention.

In addition to Amiot's interpretation, there have been four translations into Russian and at least one into German. None of the five English translations is satisfactory; even that of Lionel Giles (1910) leaves much to be desired.

Sun Tzu realized that war, 'a matter of vital importance to the State', demanded study and analysis; his is the first known attempt to formulate a rational basis for the planning and conduct of military operations. Unlike most Greek and Roman writers, Sun Tzu was not primarily interested in the elaboration of involved stratagems or in superficial and transitory techniques. His purpose was to develop a systematic treatise to guide rulers and generals in the intelligent prosecution of successful war. He believed that the skilful strategist should be able to subdue the enemy's army without engaging it, to take his cities without laying siege to them, and to overthrow his State without bloodying swords.

Sun Tzu was well aware that combat involves a great deal more than the collision of armed men. 'Numbers alone' he said, 'confer no advantage.' He considered the moral, intellectual, and circumstantial elements of war to be more important than the physical, and cautioned kings and commanders not to place reliance on sheer military power. He did not conceive war in terms of slaughter and destruction; to take all intact, or as nearly intact as possible, was the proper objective of strategy.

Sun Tzu was convinced that careful planning based on sound information of the enemy would contribute to a speedy military decision. He appreciated the effect of war on the economy and was undoubtedly the first to observe that inflated prices are an inevitable accompaniment to military operations. 'No country', he wrote, 'has ever benefited from a protracted war.'

He appreciated the decisive influence of supply on the conduct of operations, and among other factors discusses the relationship of the sovereign to his appointed commander; the moral, emotional, and intellectual qualities of the good general; organization, manoeuvre, control, terrain, and weather.

In Sun Tzu's view, the army was the instrument which delivered the *coup de grâce* to an enemy previously made vulnerable. Prior to hostilities, secret agents separated the enemy's allies from him and conducted a variety of clandestine subversive activities. Among their missions were to spread false rumours and misleading information, to corrupt and subvert officials, to create and exacerbate internal discord, and to nurture Fifth Columns. Meanwhile, spies, active at all levels, ascertained the enemy situation. On their reports, 'victorious' plans were based. Marshal Shaposhnikov was not the first to comprehend that the prerequisite to victory is 'to make proper preparations in the enemy's camp so that the result is decided beforehand'. Thus, the former Chief of the General Staff of the Red Army continues in a remarkable paraphrase of Sun Tzu, 'the victorious army attacks a demoralized and defeated enemy'.

'The Art of War' has had a profound influence throughout Chinese history and on Japanese military thought; it is the source of Mao Tse-tung's strategic theories and of the tactical doctrines of the Chinese armies. Through the Mongol-Tartars, Sun Tzu's ideas were transmitted to Russia and became a substantial part of her oriental heritage. 'The Art of War' is thus required reading for those who hope to gain a further understanding of the grand strategy of these two countries today.

S.B. GRIFFITH

# ACKNOWLEDGEMENTS

This book is a considerably revised version of a thesis submitted to Oxford University in October 1960 in part satisfaction of requirements for the Degree of Doctor of Philosophy.

In preparing it for publication I have received encouragement and advice from several friends who read and commented extensively on a preliminary draft. Notable among these is Captain B. H. Liddell Hart, to whom I am also deeply indebted for the Foreword. I wish to thank Colonel Saville T. Clark and Colonel Robert D. Heinl, U.S. Marine Corps, and Captain Robert B. Asprey for valuable critical suggestions.

I am grateful to Colonel Susuma Nishiura, Chief, War History Division, Imperial Defense Agency, Tokyo, for helping me to procure copies of various Japanese editions of *Sun Tzu*.

The final draft of the typescript was read by Professor Norman Gibbs and my Oxford tutor, Dr. Wu Shih-ch'ang, whose comments were invariably helpful. Dr. Wu's encyclopaedic knowledge of classical Chinese and of the history and literature of his native country clarified for me many constructions and allusions which would otherwise have been obscure.

I wish to thank Professor Dirk Bodde and the Princeton University Press for permission to quote from his translation of Fung Yu-lan's *History of Chinese Philosophy*; Professor Robert Hightower and the Harvard University Press for allowing me to quote from his translation of *Han Shih Wai Chuan*; Dr. Lionello Lanciotti of the University of Rome and the journal *East and West* for use of a paragraph of his scholarly essay 'Sword Casting and Related Legends of China'; and Professor C. P. Fitzgerald and the Cresset Press for permitting quotation from *China: A Short Cultural History*.

Messrs. Kegan Paul, Trench, Trubner & Company have authorized citation from Liang Ch'i-ch'ao's *Chinese Political Thought*; Dr. Homer Dubs, Professor Emeritus of Chinese, Oxford University, has allowed me to use several paragraphs from his *Hsün Tze, The Moulder of Ancient Confucianism* and *The Works of Hsün Tzu*, both published by Arthur Probsthain, London, and Dr. Arthur Waley and Messrs. George Allen & Unwin, Ltd. have given permission to reproduce a paragraph from Dr. Waley's felicitous translation of *The Analects of Confucius*. I here also express my appreciation to the editors of Imprimerie Nationale (Paris) for sanctioning quotation from the latest edition of Maspero's classic *La Chine Antique*.

On several occasions Dr. Joseph Needham of Cambridge took time from his own demanding work to enlighten me on technical matters relating to early Chinese weapons and metallurgy. He arranged for me to communicate with Drs. Kua Mo-jou and Ku Chieh-kang of the Academia Sinica, Peking. These scholars kindly answered various questions in connexion with the date of composition of 'The Art of War'.

Professor Homer Dubs and A. L. Sadler made numerous suggestions relating to the conduct of military affairs in ancient China and medieval Japan, and I gratefully acknowledge their interest in the progress of this book.

Faulty deductions and mistakes in translation are to be ascribed entirely to me.

S.B. GRIFFITH

*Norcross Lodge, Mt. Vernon, Maine, U.S.A.*

List of abbreviations of works mentioned several times in notes

| | | | |
|---|---|---|---|
| BLS | Book of Lord Shang (Duyvendak) | HFT | *Han Fei Tzu* (Liao) |
| CA | *La Chine Antique* (Maspero) | Mao | *Collected Works of Mao Tse-tung* |
| CC | Chinese Classics (Legge) | OPW | *On the Protracted War* (Mao Tse-tung) |
| CKS | *Chan Kuo Shih* (Yang K'uan) | PTSC | *Pei T'ang Shu Ch'ao* |
| Dubs | *Hsün Tzu* | San I | *Japan, A Short Cultural History* (Sansom) |
| Duy | *Tao Te Ching* (Duyvendak) | | |
| GS | *Grammata Serica Recensa* (Karlgren) | San II | *A History of Japan to 1334* (Sansom) |
| HIWC | *Han (Shu) I Wen Chih* | SC | *Shih Chi* |
| HCP | *History of Chinese Philosophy* (Fung Yü-lan) (Bodde) | TC | *Tso Chuan* |
| | | TPYL | *T'ai P'ing Yü Lan* |
| HFHD | *History of the Former Han Dynasty* (Dubs) | TT | *T'ung Tien* |
| | | WSTK | *Wei Shu T'ung K'ao* |

15

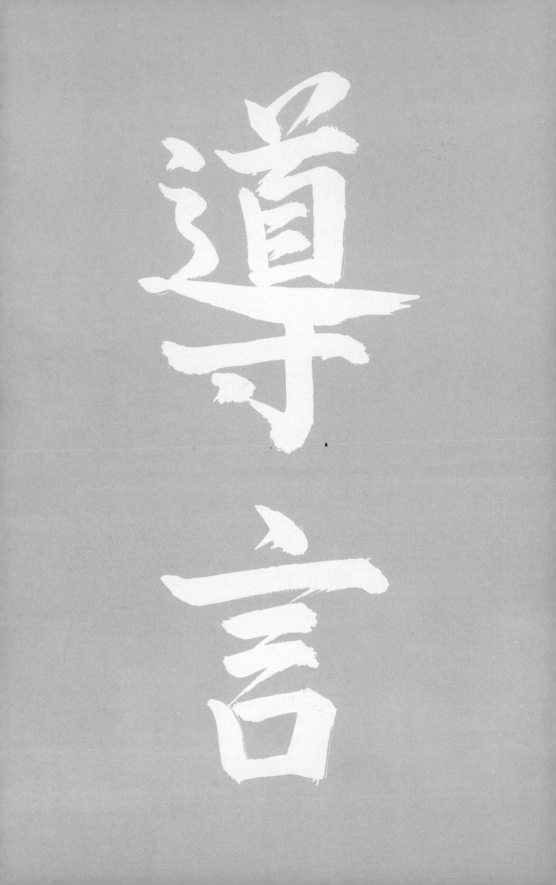

導言

# INTRODUCTION

## I
## THE AUTHOR

Over the centuries countless Chinese critics have devoted a great deal of attention to examination of literary works ascribed to the 'classical' period, an era usually defined as extending from 551 BC, the probable birth year of Confucius, to 249 BC, when King Chao of Ch'in liquidated the Chou dynasty.

One of the principal results of this scholarly endeavour has been to confirm, or more often to disprove, traditional claims relating to the authenticity of the works in question. 'The Art of War' has not escaped the careful attention of dozens of these learned analysts, who generally agree that 'The Thirteen Chapters' could not have been composed about 500 BC, as the Grand Historiographer Ssu-ma Ch'ien alleged, but belongs to a later age.

The first to doubt the reliability of Ssu-ma Ch'ien's biography of Sun Wu was an eleventh-century Sung scholar, Yeh Cheng-tsê, who concluded that Sun Wu never existed and that 'The Art of War' ascribed to him was 'probably a fabrication of disputatious sophists' of the Warring States period (453–221 BC[1]). In support of this opinion he noted that Sun Wu (who according to Ssu-ma Ch'ien was a general in Wu during the reign of King Ho-lü) was not mentioned in Tso Ch'iu-ming's commentary on the Spring

---

[1] Modern Chinese scholars consider that the period described as 'The Warring States' commenced in 453 BC with the dissolution of the kingdom of Chin. Orthodox scholarship has preferred 403 BC, the year in which King Wei Lieh of Chou legitimatized the action taken fifty years previously by the Wei, Chao, and Han Clans.

and Autumn Annals of Lu State. He also observed that armies of the Spring and Autumn (771–481 BC) were invariably command-ed by rulers, members of their families, powerful vassals, or trust-ed ministers, and that not until the period of the Warring States were they commanded by professional generals.[2] Consequently, he continues, 'the difficulty' of operational control being exercised from the capital 'did not exist during the Spring and Autumn and only began during the Warring States'. His contemporary, Mei Yao-ch'en (one of the commentators on 'The Art of War'), stated his opinion bluntly: 'This is a book of theories of the Warring States period when each antagonist tried to outwit the other.' These views were challenged by another Sung scholar, Sung Lien, who accepted the traditional biography as valid, but adduced little persuasive evidence in support of his position.

In *A Study of Apocryphal Books Ancient and Modern,* the Ch'ing critic Yao Ch'i-heng (b. AD 1647) expressed two doubts as to the authenticity of the *Sun Tzu.* First he points out (as Yeh Cheng-tsê had previously) that neither the author nor the work attributed to him is mentioned in Tso's commentary. If Sun Wu did in fact 'defeat Ch'u, enter Ying, and achieve great merit' how does it hap-pen, he asks, that the author of such 'dazzling achievements' was 'ignored' by Tso Ch'iu-ming, who 'was most detailed in respect to the affairs of Wu'?[3] He concurs with Yeh that the tale of Sun Tzu's experiment with King Ho-lü's concubines is 'fantastic' and not

---

[2] When the armies of Chin were reorganized by Duke Wen in 636 BC he assigned command of each column to a powerful vassal. In Ch'i the three columns were commanded respectively by the sovereign, the heir apparent, and the second son. When the army of King Ho-lü invaded Ch'u in 506 BC it was commanded by the First Minister, Wu Tzu-hsü. There were no titular generals until the Warring States. This critic is cited WSTK, p.939.

[3] WSTK, p.940.

worthy of belief,[4] and quotes with approval Yeh's statement relating to the exercise of military command, which, he adds, was a 'profound' remark. Finally:

> But then, did this Sun Wu exist or did he not? Did he exist, but not
> necessarily as Ssu-ma Ch'ien relates? Was the book ascribed to him
> written by him? Or was it written by one of his later disciples? None of
> this can be determined.[5]

Ch'uan Tsu-wang, another Ch'ing critic, doubted that Sun Wu was an historical personage. He concurred with Yeh Cheng-tsê that Sun Wu and his book were fabrications of 'disputatious sophists', a theory which according to him 'resolved the suspicions of a thousand generations'. As an afterthought he added: 'Naturally the Thirteen Chapters were produced by someone well versed in military matters.'[6]

Yao Nai (1732–1815) conceded that Sun Wu may have lived in Wu or at least visited that state, 'yet the Thirteen Chapters were not written by him.' They were composed later, in the period of the Warring States, 'by those who discussed military matters and were attributed to him. That is all.'[7] Liang Ch'i-ch'ao, a respected modern critic, concurs in the opinion that 'The Art of War' was written during the Warring States: 'What the book says about the pattern of war, battle tactics and planning is not possibly relevant to the Spring and Autumn.'[8]

In his *History of Chinese Philosophy* Fung Yu-lan adverts more than once to the problem of authorship of early literary works. In a discussion of Mo Tzu (479–381 BC) he writes:

> So far as we know today, the earliest work to have been composed by

---

[4] WSTK. This story is not so 'fantastic' as to be unworthy of belief. Chinese history is replete with stories every bit as fantastic. This type of criticism is not objective and can be accorded no weight.
[5] Ibid., p.941. [6] Ibid. [7] Ibid. [8] Ibid.

anybody, in a private rather than an official capacity, is the *Lun Yu*, which is a record of the most simple and abbreviated type, of Confucius' sayings. Later . . . there is a distinct advance from disjointed conversations of this kind to records of conversations of considerable length, displaying a definite story-like structure. This was the first great development in style of the writings of the Warring States philosophers. Still later such records were replaced by true essay writing. . . .[9]

Structure is thus of considerable importance in dating a Chinese work said to be ancient. The type of thematic development found in the *Sun Tzu* is first encountered in Chinese literature of the Warring States.

According to Fung Yu-lan, nobody during the Spring and Autumn 'wrote books under his own name expressing his own opinions, in contradistinction to authorship of historical works or other writings directly connected with official position'.[10] Earlier scholars had arrived at the same conclusion; Fung quotes the eminent eighteenth-century historian Chang Hsüeh-ch'eng in support of his views:

> During the early period there were no instances of the (private) writing
> of books. The officials and teachers preserved the literary records, and
> the historians made record of the passage of events. . . . It was only
> when the times were out of joint that teachers and scholars set up their
> (own private) teachings. . .[11]

Those who agree with Sung Lien that the *Shih Chi* biography of Sun Tzu is credible base their case primarily on the references in the text to the enmity between the states of Wu and Yüeh and conclude that Sun Tzu must have lived before Yüeh extinguished Wu in 474 BC. In chapter vi Sun Tzu observes that although the troops of Yüeh were 'numerous', their superiority would avail

[9] HCP i, pp.80–81.   [10] HCP i, p.7.   [11] Ibid.

them nothing 'in respect to the outcome'. Again in chapter xi, when speaking of mutual cooperation between elements of the army, he states that although the peoples of Wu and Yüeh were enemies, they would, if together in a boat tossed by the waves, co-operate 'just as the right hand does with the left'.

These references do not, however, necessarily substantiate the traditional date. They could have been deliberately inserted to mislead the reader into believing that the book was more ancient than it actually was. Such historical allusion is a technique of liter-ary forgery which flourished particularly during the Warring States when anonymous authors frequently sought to endow their works with the authority of antiquity.

The first verse in chapter xiii declares that during war 'the affairs of seven hundred thousand families are disrupted'. This suggests to the Soviet sinologue N. Konrad that the author must have lived in an era when the agricultural system known as *ching t'ien* ( 井 田 ), which Konrad identifies with a 'slave' economy, prevailed. Konrad also cites the references to Hegemonic kings in chapter xi to sup-port his theory that the classic must be assigned to the period of the *Wu Pa,* or the 'Five Lords Protector', that is, to the seventh or at latest the early sixth century BC. These are superficially the most compelling arguments in support of the traditional date, or, if Konrad's hypotheses are to be accepted, of an even earlier one.

Scholars disagree as to whether the agricultural pattern described by the term *ching t'ien* (literally, 'well field') actually existed. Some authorities believe it persisted in isolated areas from the early Chou (i.e. the late twelfth century BC) until abol-ished by Shang Yang shortly after 340 BC. Hu Shih has main-tained that it represented an agricultural Utopia dreamed up by an idyllic chronicler. Maspero, who believes it did exist, says it was abolished in Chin before that state was broken up. He writes:

> ... aussi le Tsin fut-il le premier pays où le vieux système compliqué du
> *ching* ( 井 ) disparut, remplacé par un système plus simple d'allocations
> de terres par famille, et non par groupe de huit familles, système que,
> s'il n'apportait pas encore aux paysans la propriété de la terre, était du
> moins un pas en avant très net.[12]

Mencius (398–314 BC) describes such a system, but some critics believe it to have been a product of his fertile imagination, and
one scholar has written: 'This may have been an ideal system
devised by Mencius himself. It is not likely that there could have
been such a clear-cut system.' However, Sun Tzu refers indirectly
to this pattern of agriculture, and consequently must either have
seen it in actual operation or have read about it.

In the *ching t'ien,* eight peasant families were allotted land sur-
rounding a central plot (see the character *ching)* which they cultivated for the lord of the manor.[13] The produce of the other eight,
also communally cultivated, was the property of the peasants. If
an able-bodied young man were taken from one of these households his share of work in the fields naturally would have to be
assumed by the seven households from whom none had been con-

---

[12] CA, p.267. Duyvendak thinks the system did actually exist. See BLS, pp.41–42 and note I. Professor
Dubs describes the system in HFHD (iii, pp.519–21) and states: 'The Confucian tradition was that this
system had been universal in Chou times in all flat regions and that other types of terrain had been parcelled out proportionately.' He adds that it looks excellent 'on paper', but appears to be sceptical as to
the feasibility of practical application. It is possible that the author of 'The Art of War' derived his information about the system from the works of Mencius. Both Hu Han-min and Liao Ch'ung-k'ai maintained
that the *ching t'ien* existed. An academic war, triggered off by Hu Shih, was waged for several years in
learned journals, but the entire question remains unsettled.

[13] The peasants were not slaves. They were serfs, and were strictly controlled and regimented:
Sa vie entière, publique et privée, était régie, non pour lui, mais pour toute la communauté à la fois, par
le souverain et individuellement par ses fonctionnaires. Des agents spéciaux lui commandaient chaque
année les cultures qu'il devait faire, et les temps des semailles et de la moisson; d'autres lui ordonnaient
de quitter sa maison d'hiver pour aller travailler aux champs, et de quitter les champs pour se renfermer

scripted. Thus when one hundred thousand men were mobilized, the affairs of seven hundred thousand families were 'disrupted'. But if the system did indeed exist for hundreds of years prior to about 350 BC it is obvious that general references to it cannot be of much value in the precise dating of documents.

Nor does use of the term 'Hegemonic king' necessarily indicate that 'The Art of War' was composed in the historical period of 'The Lords Protector' *(Wu Pa)*. As late as *c.* 250 BC we find Han Fei Tzu saying that the enterprise 'of the Hegemonic Ruler is the highest goal of the Lord of Men'.[14] The term 'Hegemonic king' is used in 'The Art of War' in the sense Han Fei Tzu used 'Hegemonic Ruler'.

In ancient China war had been regarded as a knightly contest. As such, it had been governed by a code to which both sides generally adhered. Many illustrations of this are found in the Tso Chuan. For example, in 632 BC the Chin commander, after defeating Ch'u at Ch'eng P'u, gave the vanquished enemy three days' supply of food. This courtesy was later reciprocated by a Ch'u army victorious at Pi. By the time 'The Art of War' was written this code had been long abandoned.

During the Spring and Autumn armies were small, inefficiently organized, usually ineptly led, poorly equipped, badly trained, and haphazardly supplied. Many campaigns ended in disaster simply because the troops could find nothing to eat. The invasion of Ch'u by Wu in 506 BC which culminated in the capture and

---

dans sa maison; d'autres encore s'occupaient de son mariage; d'autres lotissaient les terrains et distribuaient les parts supplémentaires suivant le nombre des enfants. (CA, p.95).
In further refutation of Konrad's 'slave economy' hypothesis it may be added that Sun Tzu's solicitude for the welfare of the people and his realization that their morale and that of the army must be sustained if war is to be successfully prosecuted are inconsistent with a society in which the mode of production was based on the institution of slavery.

[14] HFT ii, p.240.

destruction of Ying, the Ch'u capital, is one of the rare examples of a successful lengthy campaign in the entire Spring and Autumn period, when issues were ordinarily settled in a day.[15] Of course, cities were besieged and armies sometimes kept in the field for protracted periods. But such operations were not normal, first because they were impractical and second because keeping the army over the season was morally interdicted.

The author of 'The Art of War' lived at a time when large armies were effectively organized, well trained, and commanded by professional generals. In the opening verse of chapter ii the phrase 'one hundred thousand armoured troops' is used in a discussion of the related problems of war finances, supply, and replenishment. Armies of this size were unknown in China before 500 BC.

The armies Sun Tzu discusses were composed of tactical elements capable of independent and co-ordinated manoeuvre and responsive to control exercised by use of bells, gongs, drums, flags, and banners. The undrilled peasant levies of the Spring and Autumn could not possibly have been capable of such manoeuvres.

Sun Tzu's definitions of the qualities to be sought in a good general indicate that he did not conceive high military office to be a prerogative limited to an hereditary aristocracy as was the earlier custom. His concern with the relationship between a field commander and the sovereign reflects his interest in establishing the authority of the professional general. In chapter iii he names the ways in which a ruler may bring misfortune upon the army by interfering with its operations and administration. And in chapter viii he asserts that the general, having received the mandate of

---

[15] It is impossible to make dogmatic statements about early Chinese military history. Campaigns were frequently interrupted by rebellion at home, by an attempted *coup d'état*, or a sudden attack launched during the absence of the army.

command, is not required to obey the orders of the sovereign blindly, but should act as circumstances dictate. This concept is completely inconsistent with traditional thought.

In chapter xiii Sun Tzu describes the organization, financing, and direction of secret operations. Here he observes that what enables the enlightened ruler and the sage general to achieve results beyond the capability of the ordinary man is foreknowledge. This he asserts cannot be elicited from disembodied spirits, supernatural beings, or analogy with past events, but only from men who know the enemy situation. His adjuration against divination and the taking of omens (which he says must be rigorously prohibited) is incompatible with the Spring and Autumn when belief in spirits was universal and divination by tortoise-shell and yarrow-stalks an essential preliminary to any undertaking which involved the fortunes of a ruling house.[16]

Sun Tzu's theory of war and strategy and his tactical doctrine are of course germane to the problem of dating. Because war is of vital importance to the state and a question of survival or ruin it must be thoroughly investigated. Thus, the opening verses of the classic outline a method for analysis of the factors which constitute military strength. This process, now described as an 'Estimate (or Appreciation) of the Situation', is rational and is scarcely consistent with the mentality characteristic of the Spring and Autumn, when rulers indulged in military adventures to satisfy a whim, to revenge a slight or an insult, or to collect booty.

When Sun Tzu wrote, war had become a dangerous business; the recourse when other means had failed. The best policy, he

---

[16] Confucius had earlier expressed his scepticism when questioned about the spirit world. Apparently he was one of the first to do so.

says, is 'to attack the enemy's plans'; the next best to disrupt his alliances, for 'to subdue the enemy's army without fighting is the acme of skill'. This indicates his perception that war was no longer a regulated pastime, but the ultimate instrument of statecraft.

The strategic and tactical doctrines expounded in 'The Art of War' are based on deception, the creation of false appearances to mystify and delude the enemy, the indirect approach, ready adaptability to the enemy situation, flexible and coordinated manoeuvre of separate combat elements, and speedy concentration against points of weakness. Successful application of such tactics requires highly mobile and well-trained shock and *élite* troops. Such formations were not common until the Warring States.

In chapter ii, and again in chapter v, there are specific references to the crossbow. The first reads: 'As to government expenditures, those due to broken-down chariots, worn-out horses, armour and helmets, arrows and crossbows . . .'; the latter: 'His potential is that of a fully drawn crossbow; his timing the release of the trigger.' Precisely when the crossbow, which revolutionized Chinese warfare, was introduced has not been determined; most scholars put the date at about 400 BC. The first reference to it is found in Ssu-ma Ch'ien's description of the battle at Ma Ling fought in 341 BC where Sun Pin, Chief Strategist of Ch'i, defeated a Wei army commanded by his erstwhile friend P'ang Chüan. The ten thousand crossbowmen Sun Pin had placed in ambush practically annihilated the enemy. Again in chapter xi Sun Tzu uses 'release of the trigger' as a figure of speech to describe the sudden unleashing of an army's potential energy.

The term *chin* as a generic word descriptive of 'money' or 'metallic currency' came into use during the Warring States.[17] Although money was cast in several forms during the late Spring and Autumn its acceptance as a medium of exchange was inevitably a gradual process. Certainly a term specifically descriptive of metallic currency would not have been used five times in the text unless money was common.

The phrase meaning 'armoured troops', or 'troops wearing armour', occurs in the first verse of chapter ii.[18] There were no 'armoured troops' in the Spring and Autumn. At that time only the *shih*[19]—the chariot-riding nobility—and their immediate retainers carried primitive shields of lacquered leather or varnished rhinoceros hide. The footmen wore padded jackets; only much later were they to be provided with protective garments of treated sharkskin or animal hide.

Sun Tzu uses the character *chu* in the sense of 'sovereign' eleven times.[20] In the Spring and Autumn this character meant 'lord' or 'master' and was used in addressing a minister. The connotation 'sovereign' was a later assimilation. This anachronism was noticed by the Ch'ing scholar Yao Nai.

In chapter vii we read: 'In a forced march of fifty *li,* the commander of the *Shang* (Upper, or Van) army will be captured.' The terms *Shang Chiang, Chung Chiang,* and *Hsia Chiang* to designate the generals in command of the traditional 'Three Armies' were not commonly used until the Warring States.[21] Two terms which occur in chapter xiii, 'Secret Operations', did not have the specific meaning in which they are used in the context until the Warring States. These are *yeh che,* meaning 'chamberlain', 'receptionist', or 'usher', and *she jen,* meaning 'retainers' or 'bodyguard'.[22]

---

[17] CKS, p.9. The character is *chin* (金) [18] *Tai Chia* (帶甲) [19] Only the *shih* (士) were permitted to ride chariots to battle, a prerogative they enjoyed as well in time of peace. [20] *Chu* (主) [21] CKS, p.9. [22] *Yeh Che* (謁者) ; *she jen* (舍人)

Sun Tzu believed that the only constant in war is constant change and to illustrate this he used several figures of speech, among which is 'of the five elements, none is always predominant'. The theory of the constant mutation of the five 'powers' or 'elements': earth, wood, fire, metal, and water, did not develop as a philosophical concept until the Warring States. In earlier times they appear to have been regarded as the five elemental substances.

It is significant that Sun Tzu does not refer to cavalry. Cavalry was not made an integral branch in any Chinese army until 320 BC when King Wu Ling of Chao State introduced it—and trousers. It is reasonable to assume that if cavalry had been familiar to Sun Tzu he would have mentioned it. This is interesting negative testimony that 'The Art of War' was not written in the third century BC as Maspero believed.[23]

Thus in respect to date of composition, we derive from the text itself, the best possible source, evidence which indicates almost beyond a doubt that the work was written at least a century (and more likely a century and a half) after Ssu-ma Ch'ien says it was. We may therefore assign this first of all military classics to the period *c.* 400–320 BC.

What, then, is the source of the Sun Tzu legend which the Grand Historiographer perpetuated? What is the explanation of the connexion between this Sun Tzu of myth and Wu State? Professor Ku Chieh-kang has proposed the following ingenious theory:

> It can be presumed that when Ch'i in 341 BC launched a punitive expe-
> dition against Wei in order to relieve Han, T'ien Chi was the general
> and Sun Pin was the strategist. T'ien Chi later on fled to Ch'u and

---

[23] Maspero suspects that 'The Art of War'—which he describes as 'un petit opuscule' (CA, p.328) and 'ce petit ouvrage' (ibid., note I)—is a forgery of the third century BC and cannot properly be ascribed either to Sun Pin or his 'fabulous ancestor'.

Ch'u made T'ien Chi a feudal lord in Chiangnan which was Wu territory. It might be that Sun Pin had followed T'ien Chi to the Chiangnan area and there wrote his *Sun Tzu Ping Fa*. Later the people made a chronological mistake and described him as having lived during the Spring and Autumn period and, furthermore, created a Sun Wu who helped Ho-lu in his invasion of Ch'u, and this story was adopted by Ssu-ma Ch'ien.[24]

If Sun Pin did write 'The Thirteen Chapters' it would not in his time have been considered unethical to ascribe the work to a figure alleged to have lived a century and a half before. Or did 'people' simply make a 'chronological mistake' as Professor Ku Chieh-kang opines? Doctor Kuo Mo-jo writes: 'The biography of Sun Wu is not dependable; it is fiction. The *Sun Tzu Ping Fa* was written during the Warring Kingdoms period; its authorship is unknown. It is difficult to determine whether Sun Pin was the author.'[25]

Possibly this work is a compendium of the teachings of an unknown Warring States strategist. *The Analects* of Confucius is just such a compilation. As Fung Yu-lan points out, the question of authorship of ancient literary works is one which frequently defies solution:

The conception of authorship was evidently not wholly clear in early China, so that when we find a book named after a certain man of the Warring States period, or earlier, this does not necessarily mean that the book was originally actually written by that man himself. What part of it was the addition of his followers, and what part was by the original author, was not at that time looked upon as requiring any distinction and hence today cannot for the most part be distinguished any longer.[26]

---

[24] Personal correspondence. [25] Personal correspondence. [26] See following page.

29

Thus we arrive finally at the same impasse as did Yao Ch'i-heng three centuries ago. We do not know if this Sun Wu existed; we do not know if the work ascribed to him was written by him, and we are therefore forced, with the eminent Ch'ing scholar, to place the *Sun Tzu* in the category of 'Authorship Unsettled'. But the originality, the consistent style, and the thematic development suggest that 'The Thirteen Chapters' is not a compilation, but was written by a singularly imaginative individual who had considerable practical experience in war.

## II
## THE TEXT

Is the present text of Sun Tzu's classic identical with 'The Thirteen Chapters' familiar to Ssu-ma Ch'ien? To this tantalizing question there can be no conclusive answer, for the early records relating to 'The Art of War' are slightly confused.

In the late first century BC the scholar Liu Hsiang was directed by the Emperor to begin collecting literary works for the Imperial Library. After his death this task was continued by his son, Liu Hsin.[1] In his 'Seven Syllabi' Liu Hsiang had noted the existence of an 'Art of War' by one Sun Tzu 'in three rolls', and some years later Pan Ku,[2] the historian of the Former Han dynasty, listed 'The Art

---

[26] (Previous page.) HCP, p.20. But Sun Hsing-yen, whose edition of the classic has been considered standard for almost two centuries, was convinced that the work had been written by Sun Tzu: 'The words of the philosophers were all put down after they departed this world by followers and pupils who recorded them and made them into books. Only this book was written by Sun Tzu's own hand. Moreover, it antedates the *Lieh Tzu, Chuang Tzu, Meng Tzu* and *Hsün Tzu* and is really an old book.' (WSTK, p.941. These works are said to have been composed, respectively, in the third century BC; between 369 and 286 (?) BC; between 372 and 289 (?) BC, and in the mid-third century BC.)
[1] 53 BC–AD 23. [2] AD 32–92.

of War of Sun Tzu of Wu in eighty-two chapters with nine rolls of diagrams' as being in the Imperial Library.[3] Pan Ku assigned this compilation to Jen Hung, who as a colleague of Liu Hsiang had catalogued military works. But in the preface to his edition, probably composed about AD 200, Ts'ao Ts'ao refers to thirteen chapters. Sixty-nine chapters thus mysteriously appeared and as mysteriously vanished between the time Ssu-ma Ch'ien wrote the *Shih Chi (c.* 100 BC) and Ts'ao Ts'ao wrote his 'Brief Explanation' about three hundred years later. Or did they? Did these sixty-nine chapters ever in fact exist? I suspect they did not.

There are several possible explanations for this disparity in figures. One is that attributed material and commentary had become associated with the text during the interval between the composition of the *Shih Chi* and the record of eighty-two chapters ascribed by Pan Ku to Jen Hung. It may be that Jen Hung assembled all the material relating to Sun Tzu without attempting to distinguish the original text from accretions.

Or the difference in figures may be due to the method by which books were manufactured, or put together, in ancient China. At that time paper had not yet been invented and records were ordinarily written with ink made of soot on sections of bamboo or on thin wooden slips. A wooden slip or a split of bamboo suitable for this purpose would have measured 8 to 10 inches in length and about ³/₄ inch in width. Twelve to fifteen characters (depending of course on individual calligraphy) were written on these slips, many of which have been found. The edges of the slips were sometimes notched or pierced in two places so that they could be connected sequentially by thongs made of leather, silk, or

---

[3] See HIWC.

hemp.[4] Sometimes slips were perforated at one end and strung together. When the first system was used chapters or sections of books could have been rolled for storage exactly as a scroll is today.

As the text of the *Sun Tzu* runs to something over thirteen thousand characters, approximately one thousand such slips would have been needed to record it. Had these been assembled as a unit the resulting 'book', when spread, would measure over 60 feet in length, and when rolled would have required an ox-cart to transport. But books were divided then, as they are today, into sections or chapters. These sections were naturally of varying length; some may have required a dozen slips, others twenty or more. This suggests the possibility that Jen Hung or a copyist made a careless mistake and recorded eighty-two 'rolls' as eighty-two 'chapters'. Errors in transcription were by no means uncommon. But how do we reconcile the three rolls mentioned in the 'Seven Syllabi' with the figure 'eighty-two'? Only by presuming that the 'Seven Syllabi' text was written on silk rather than on slips.

Tu Mu[5] attempted to solve this problem by accusing Ts'ao Ts'ao of condensing the text by 'pruning away redundancies'. Indeed Ts'ao Ts'ao had laid himself open to the charge, for in his introduction he says that as the texts current in his time 'missed the essential meaning' he had undertaken to write a 'brief explanation'. Thus it appears that Ts'ao Ts'ao may have edited the text and excised portions he considered to be accretions. Probably Ts'ao Ts'ao had various versions available and by comparison of them produced what he considered to be an authentic text. It is this, with his commentary, which has been preserved. And while we cannot know

---

[4] This accounts for the confused state of some of the ancient works when they were rediscovered. The thongs or strings rotted and the slips fell apart.

[5] AD 803–52.

that it is identical with the *Sun Tzu* familiar to Ssu-ma Ch'ien, I think it reasonable to assume that in essential respects it is.

As this work was well known in the late fourth century BC it must already have been in circulation for a number of years. In chapter iii of 'The Book of Lord Shang' at least half a dozen of Sun Tzu's verses are paraphrased.[6] While it is true that the book ascribed to Shang Yang (who was torn apart by chariots in 338 BC) was probably compiled by members of the Legalist School shortly after his death, scholars believe it reflects his sayings and opinions.

Han Fei Tzu, another Legalist statesman and author, who died toward the end of the third century BC, was acquainted with the *Sun Tzu;* in the chapter of his work entitled 'Five Vermin' he remarks that 'in every family there are men who preserve copies of the books of Sun Wu and Wu Ch'i'.[7] In 'The Way to Maintain the State' he observes that in a Legalist Utopia 'the tactics of Sun Wu and Wu Ch'i would be abandoned', presumably because with 'All-under Heaven' brought under the centralized totalitarian rule he advocated there would be no occasion for war.

Hsün Tzu (*c.* 320–235 BC) also refers to both Sun Tzu and Wu Ch'i. In 'A Debate on Military Affairs', there is recorded a discussion between the philosopher and General Lin-wu in the presence of King Hsiao-ch'en of Chao.[8] The general had obviously studied 'The Thirteen Chapters' thoroughly; the line of his argument follows Sun Tzu almost to the letter:

> What is valuable in military affairs is strength and advantage; what is done is sudden alteration of troop movements and deceitful stratagems. He who knows best how to manage an army is sudden in his

---

[6] BLS, pp.244–52.
[7] HFT ii, p.290.
[8] Reigned 265–245 BC.

movements; his plans are very deep-laid, and no one knows whence he may attack. When Sun and Wu led armies, they had no enemies in the whole country.[9]

Hsün Tzu, a Confucian, argues from a moral and ethical basis, and as one might expect, demolishes his opponent: 'The armies of the benevolent *(jen)* man cannot use deceit.'[10] But Hsün Tzu was not impractical: he approved of armies 'for the purpose of stopping tyranny and getting rid of injury'.[11]

Shortly after unifying China in 221 BC the First Universal Emperor, Shih Huang Ti, ordered the collection and destruction of books which his Legalist advisers considered pernicious. This edict was aimed specifically at works attributed to Confucius and his school. Certain types of books such as those dealing with technical subjects were declared exempt, and as Ch'in was a wholly militarized state (and had been one for a century) it is logical to assume that works which related to the art of war would have been spared. (This book-burning was not in fact the complete holocaust which later Confucian scholars represent it to have been, for apparently the edict was none too strenuously enforced. It was revoked in 196 BC by the Han Emperor Hui. The later Emperor Wu offered generous rewards to all those who presented copies of ancient works.)

In 81 BC a number of scholars were summoned to the capital for discussions. One of the important matters on the agenda was how the administration of the Empire might be improved. The literati unanimously agreed that government monopolies in salt, iron, and fermented liquor should be abolished. This provoked a great debate at which the Emperor Chao presided.[12] Some time

---

[9] Dubs, i, p.158. [10] Ibid., p.159. [11] Ibid., p.161.
[12] Trans. Esson M. Gale as 'Debate on Salt and Iron'.

later Huan K'uan recorded the substance of the arguments in which appear one direct quotation from 'The Art of War' and paraphrases of several verses. This is striking testimony to the esteem in which 'The Thirteen Chapters' were then held.

There is no record of Han commentators on Sun Tzu's work, but there must have been some, for as early as Liu Hsiang's time there were sixty-three recognized 'schools' of military theoreticians. In AD 23 the usurper Wang Mang 'appointed to office the [various] persons skilled in the methods of the sixty-three schools of military arts whom he had summoned.'[13]

Toward the close of the second century AD and in the early years of the third we find commentators of record. In addition to Ts'ao Ts'ao, Wang Ling, known as 'Master Wang', Chia Yeh, and Chang Tzu-shang, wrote works no longer extant. During the Liang dynasty (AD 502–56) at least two editions of Sun Tzu appeared. The interpretations of a 'Mr.' Meng are preserved in the 'Ten School' version; Shan Yu's have been lost.

Among T'ang (618–905) critics, Tu Yü, his grandson the poet Tu Mu, Li Ch'üan, Ch'en Hao, and Chia Lin were the most respected. Separate editions containing the interpretations of the last four were published; Tu Yü's comments were included with the text of 'The Thirteen Chapters' contained in his monumental encyclopedia, the *T'ung Tien* (chapters 148–63). Selected verses from Sun Tzu were included in other anthologies completed during the T'ang, notably in the *Pei T'ang Shu Ch'ao* (chapters 115–16).

Interest grew in the Sung dynasty; 'The Thirteen Chapters' were included in the *T'ai P'ing Yü Lan* (chapters 270–337). Of

---

[13] HFHD iii, p.442 n.213.

numerous published commentaries, those of Mei Yao-ch'en, Wang Hsi, Ho Yen-hsi, and Chang Yü were selected by both Chi T'ien-pao and Cheng Yu-hsien for inclusion in their editions, respectively entitled *Shih Chia Chu* ('Ten School') and *Shih I Chia Chu* ('Eleven School').[14] Both also included Ts'ao Ts'ao and 'Mr.' Meng as well as the four T'ang scholars previously named. In both collations comments were arranged in chronological order under the verses to which they apply. As Tu Yu's comments had been included in his *T'ung Tien* rather than in a separately published work, Chi T'ien-pao did not consider that he represented a 'school'. The difference in the titles of these two compilations is therefore of no importance. Chi T'ien-pao's edition was later incorporated into The Taoist Canon; it was this which Sun Hsing-yen copied when he discovered it at a Taoist monastery in Shensi province.

But it was not the individual works of these various scholars, important as they were, which gave the greatest impetus to study of 'The Art of War' during the Sung. This derived rather from an edict issued by the Emperor Sheng-tsung (1068–85) which designated seven 'Martial Classics' by name and prescribed them as obligatory study for aspirants to commission in the army. Ts'ao Ts'ao's edition was placed first on this list.[15]

The edict further directed that cadets pursue their studies under the direct supervision of a *Po Shih,* a scholar of high rank. The *Po Shih* appointed to this important post was Ho Chu-fei, who, as first superintendent of the Imperial Military Academy, chose as the basic text for his young students an edition of Sun Tzu with 'Lecture Notes' by the well-known critic Shih Tzu-mei.

---

[14] 十家註; 十一家註.

[15] Ts'ao Ts'ao has for centuries been regarded as one of the outstanding masters of the military art.

These 'Lecture Notes' in the form of a commentary on Sun Tzu's verses have fortunately been preserved.

There is record of but one new edition of the *Sun Tzu* during the Yüan (Mongol) dynasty (1206–1367), and this, edited by P'an Yen-wang, has been lost. But after the Mongols had been driven out of Peking and the Ming (1368–1628) firmly established, interrupted work on the classics was resumed with great vigour; during this period of three centuries over fifty commentaries, interpretative studies, and critical essays were devoted to 'The Art of War'. Of these, the most popular was the work of Chao Pen-hsüeh, which has been repeatedly reissued.

Sun Hsing-yen was the leading authority on Sun Tzu to emerge during the Ch'ing (Manchu) dynasty; it is his edition (in the preparation of which his friend Wu Jen-chi collaborated) which has been considered standard for almost two hundred years, and which is the basis for the present translation.

# III
# THE WARRING STATES

Confucius, the first and ultimately the most influential of China's philosophers, died in 479 BC. Almost exactly a quarter of a century later the leaders of the Wei, Han, and Chao Clans attacked the ruler of Chin, defeated him at Ching Yang in 453, and parcelled out his domain among themselves.[1] Earl Chih was promptly decapitated, his family exterminated, and his skull, suitably embellished, presented to Wu Hsu of the Chao, who used it as a drinking cup. Thus inauspiciously began the age of the Warring States.

---

[1] Ching Yang was on or near the site of modern T'aiyuan.

There was at this time a Son of Heaven, a member of the Chou house, whose authority outside the borders of the tiny enclave allowed him by the predatory rulers of the great states was non-existent. Indeed, for several centuries the kings of Chou had been purely symbolic figures whose principal functions were to regulate the calendar and perform the periodic ritual sacrifices upon which depended the harmonious relationship of Heaven, Earth, and Man.

The decline of the royal house had commenced with the transfer of the Chou capital from Shensi to the east in 770 BC. Its feebleness a century and a half later is made strikingly evident in the formula used by the King to invest Duke Wen of Chin as Lord Protector (*Pa*):

> Oh, my Uncle! Illustrious were the Kings Wen and Wu; they knew how to take care of their shining virtue, which rose with splendour on High (toward Heaven) and whose renown spread wide on earth. That is why the Sovereign of On-High made the Mandate succeed in the case of the Kings Wen and Wu. Have pity on me! Cause me to continue (the line of my ancestors); Me, the Unique Man and cause (me and my line) to be perpetually on the throne![2]

As of 450 BC there were eight large states in China, but Yen in the north and Yüeh in the east played no decisive part in the wars which raged almost without ceasing for the next two and a quarter centuries. The 'Big Six' were Ch'i, Ch'u, Ch'in, and 'The Three Chins'—Wei, Han, and Chao. Additionally there were a dozen smaller principalities, all of which were to be absorbed by the larger states, who ate them up as systematically 'as silkworms eat mulberry leaves'. In 447 BC Ch'u extinguished Ts'ai, a small state in today's Honan province; two years later she swallowed Ch'i (not

---

[2] Granet, pp.25–26.

the Ch'i mentioned above), and in 431 Chu. This process accelerated after 414 BC; in the next sixty-five years half a dozen small states disappeared from the historical scene. Occasionally the rulers managed to arrange recesses from the endemic wars which were produced by their insatiable ambitions. Such breathing spaces were necessary if for no other reason than to train the peasant armies periodically cut to pieces. It is extremely unlikely that many generals died in bed during the hundred and fifty years between 450 and 300 BC.

This was one of the most chaotic periods in China's long history. The forested hills, the reed-bordered lakes, the many swamps and marshes provided hiding places for the bands of robbers and cut-throats who raided villages, kidnapped travellers, and exacted toll from merchants unlucky enough to fall into their hands. Many of these outlaws were peasants who had been forced into brigandage to survive. Others were escaped criminals, deserters from the army, and disgraced officials. Altogether they constituted a formidable challenge to the so-called forces of law and order. The vendettas of the great families were conducted by bands of professional swordsmen recruited from the lower ranks of a dissolving hereditary aristocracy.

Some individuals took vigorous exception to the amoral standards of the times. The most important of these was Mo Ti, or Mo Tzu (c. 479–381 BC), who denounced the crime and futility of the wars to which the rulers of his age devoted their energies. 'Suppose', he said, 'soldier hosts arise':

> If it is in winter it will be too cold, and if in summer it will be too hot.
> So it should be done neither in winter nor in summer. But if it is in
> spring it will take people away from sowing and planting, and if in fall
> it will take them from reaping and harvesting. Should they be taken

away in any of these seasons, innumerable people would die of hunger and cold. And, when the army sets out, the bamboo, arrows, plumed standards, house tents, armor, shields, and sword hilts will break and rot in innumerable quantities and never come back. Again with the spears, lances, swords, poniards, chariots and carts: these will break and rot in innumerable quantities and never come back. Innumerable horses and oxen will start out fat and come back lean, or will die and never come back at all. An innumerable people will die because their food will be cut off and cannot be supplied on account of the great distance of the roads, while other innumerable people will get sick and die from the constant danger, the irregularities of eating and drinking, and the extremes of hunger and over-eating. Then the army will be lost in large numbers or in its entirety; in either case the number will be innumerable.[3]

Mo Ti condemned aggressive war in uncompromising terms:

If a man kills an innocent man, steals his clothing and his spear and sword, his offence is graver than breaking into a stable and stealing an ox or a horse. The injury is greater, the offence is graver, and the crime of a higher degree. Any man of sense knows that it is wrong, knows that it is unrighteous. But when murder is committed in attacking a country it is not considered wrong; it is applauded and called righteous. Can this be considered as knowing what is righteous and what is unrighteous? When one man kills another man it is considered unrighteous and he is punished by death. Then by the same sign when a man kills ten others, his crime will be ten times greater, and should be punished by death ten times. Similarly one who kills a hundred men should be punished a hundred times more heavily. . . . If a man calls black black if it is seen on a small scale, but calls black white when it is

---

[3] HCP, pp.94–95.

seen on a large scale, then he is one who cannot tell black from white. .
. . Similarly if a small crime is considered crime, but a big crime such as
attacking another country is applauded as a righteous act, can this be
said to be knowing the difference between righteous and unrighteous ?[4]

This philosophy was not particularly popular with Warring
States rulers who were actuated by the imperatives of power rather
than by the adjurations of moralists. But these sovereigns were by
no means uncivilized. Most of them were educated men who lived
luxuriously. In their courts boredom was banished by well-
equipped harems, teams of dancing girls, musicians, acrobats, and
the expert chefs who staffed the kitchens. The itinerant sophists
who with their disciples roamed from one capital to another, the
more successful in carriages, those less so on foot, were welcomed
and entertained at these courts.

Markets flourished, and in spite of chronic disorder, interstate
commerce as well. Traders and merchants made fat profits. In Lin
Tzu, the capital of Ch'i, there were in the early fourth century BC
seventy thousand households. If ten 'mouths' (a reasonable figure)
be allowed per household, the population must have been almost
three-quarters of a million. This city, then the largest and probably the richest in Chin, derived its wealth from trade. There those
who came to buy salt, silk, iron, and dried fish found entertainment in restaurants, music halls, and brothels and gambled on
dog races, cock fights, and football games. But the inarticulate
peasants, who comprised probably 90 per cent of the population,
enjoyed none of these amenities or luxuries. Their lot was toil and
war, to labour in the fields, to obey their superiors, and to keep
their mouths shut.

---

[4] Cited by Liang Ch'i Ch'ao, p.97.

This society was controlled by a legal code of frightening severity. Several thousand crimes were punishable by death or mutilation. Castration, branding, slicing off the nose, chopping off the toes or feet, cutting leg tendons, or breaking knee caps were commonly inflicted. Nor were such punishments restricted to the lower classes of society; great officers sometimes suffered them. Thus there was at least in theory equitable administration of the law. But those whose feet had been severed undoubtedly derived small consolation from contemplation of the purely academic aspects of this Draconian code.

The political environment gave ample scope to the talents of self-styled experts in every field and particularly to professional strategists. Between 450 and 300 BC successive generations were decimated with methodical regularity and war became a 'fundamental occupation'.[5] The pretence of conforming to the idyllic code of morality reputedly exemplified in the reigns of the Sage kings had long since been abandoned. Diplomacy was based on bribery, fraud, and deceit. Espionage and intrigue flourished. Nor was treasonable conduct regarded as abnormal by ambitious generals who changed allegiance or ministers of state who were willingly corrupted.

The career of the famous soldier Wu Ch'i reflects the standards of his time. Born and raised in Wei State he sought military office in Lu. To allay the Duke's doubts of his loyalty, he killed his wife because she was a native of Wei. Some time thereafter jealous officials brought false accusations against him. He caused a number of them to be murdered and fled to Wey, where he offered his

---

[5] This expression was given currency by Shang Yang (Lord Shang), Prime Minister to Duke Hsiao of Ch'in. He conceived the two 'fundamental Occupations' to be war and agriculture. He is universally execrated by traditional historians.

services to the Marquis. The Prime Minister described him as covetous and addicted to debauchery but an expert general. The Marquis hired him and made him Protector of the West River. Later he left Wey and was employed by King Tao of Ch'u, who appointed him Prime Minister in 384 BC. He reorganized and modernized the state administration, thereby gaining many enemies. When King Tao was assassinated in 381 BC Wu Ch'i was executed.[6]

This dynamic age demanded practical solutions to the problems of politics and war, and hundreds of scholars who wandered from one state to another were eager to peddle ideas to rulers 'anxious over the perilous condition of their countries and the weakness of their armies'. Sovereigns competed for the advice of battalions of professional talkers, who, in 'interminable discussions', captivated kings, dukes, and great men with arguments of 'confusing diversity'.[7] These itinerant Machiavellis were intellectual gamblers. When their advice turned out to be good they frequently attained high position; if poor, they were unceremoniously pickled, sawn in half, boiled, minced, or torn apart by chariots.

But rewards were sufficiently lucrative to induce many men to devote their talents to government, diplomacy, and military affairs, and sovereigns whose ambition was to 'roll up All-under-Heaven like a mat' and 'tie up the four seas in a bag' gave them sympathetic hearing.

> The corrosive influence of this new class upon feudal ideas and institutions was of the first importance. The wandering scholars were bound by no lasting loyalties, were attached by no sentiment of patriotism to the states they served and were not restricted by any feeling of ancient

---

[6] Wu Ch'i was born in 魏 (Wei) and fled to 隗, here, to avoid confusion, romanized 'Wey').
[7] BLS, p.95.

chivalry. They proposed and carried out schemes of the blackest treachery. Frequently they secretly served two princes at once, playing off the
policy of the one against the other. Moving from kingdom to kingdom, always with some eloquent and intricate scheme to propose, they
fought the particularism of the old feudal aristocracy, and envisaged
plans to reduce the whole empire to the obedience of the sovereign
they served. This was the bait they held out to their temporary masters;
it was no longer hegemony, but empire, which had become the aim of
state policy.[8]

The desires of these princes were identical with those of whom
Plutarch wrote that they merely made use of the words peace and
war 'like current coin to serve their occasions as expediency sug-
gested'. These occasions could be served in the China of that time
only by intrigue or war. And war, an integral part of the power
politics of the age, had become 'a matter of vital importance to the
state, the province of life or death, the road to survival or ruin'. To
be waged successfully, it required a coherent strategic and tactical
theory and a practical doctrine governing intelligence, planning,
command, operational, and administrative procedures. The
author of 'The Thirteen Chapters' was the first man to provide
such a theory and such a doctrine.

The trend toward the growth of large states at the expense of
their smaller and weaker neighbours was a constant feature of
Chinese historical development and was to culminate in 221 BC in
the establishment of a monolithic state by the First Universal
Emperor.

During the period with which we are concerned, many factors
contributed to the concentration of political power in the hands

---

[8] Fitzgerald, p.70.

of an ever-decreasing number of rulers. Of these, the development of an iron technology was perhaps of unique importance.

Iron was known in China before 500 BC, but it was extremely rare and valuable because controlled methods of smelting had not yet been devised. The swords of King Ho-lü of Wu were famous in Chinese legend; renowned Japanese sword makers of later ages traced the origins of their craft directly to Kan Chiang, Mo Yeh and their son Ch'ih Pi, whose lives, art, and the blades they forged have provided material for many folk tales and more recently for scholarly essays on the subject of iron founding in ancient China.[9] The forging of these fabulous blades was accompanied by symbolic human sacrifice; the process was ritualistic and known only to the initiated.

> Kan Chiang, from the country of Wu, had the same teacher as Mo Yeh. Together they made swords. Ho-lü, King of Wu, ordered two swords to be made; one was called Kan Chiang and the other Mo Yeh. Mo Yeh was the wife of Kan Chiang. Kan Chiang therefore made the sword after collecting the iron in the five mountains and the gold (chin?) in the six directions. He examined the sky and the earth; yin and yang shone together. A hundred *shen* came down to observe; the steam came down; but the iron and the gold did not melt. Kan Chiang could not understand the reason. Mo Yeh said to him: 'You are a good founder and the King orders you to make swords. Three months have gone by, and you cannot manage to finish them; what are your ideas on the matter?' Kan Chiang answered: 'I do not understand the reason.' Mo Yeh said: 'Then the transformation of divine (*shen*) things is in need of a person to carry it out. Now then, in making the swords what person

---

[9] See Doctor Lionello Lanciotti's scholarly and fascinating essay 'Sword Casting and related Legends in China' in *East and West* (Year VI, no.2 and no.4, of July 1955 and January 1956 respectively).

is needed for achieving the work?' Then Kan Chiang said: 'In ancient times my Master cast the metal. The gold and the iron did not melt and he with his wife entered the furnace; then it was done. Until now the succeeding (founders) have ascended the mountains, making the casting with perfumed robes of white hempen cloth, daring then to throw the gold in the furnace. Now I make the sword on the mountain and no such change takes place.' Mo Yeh answered: 'The Master knew the need of melting a body in order to achieve the work. What difficulty have I (to sacrifice myself)?' Then Kan Chiang's wife cut off her hair, and cut her nails, and threw them in the furnace. She ordered three hundred boys and maidens to work the bellows, and the gold and iron melted to make the swords. The male sword was called Kan Chiang and the female sword Mo Yeh. . . . Kan Chiang hid the male sword and took the female one and presented it to the King.[10]

It was not, however, for almost one hundred years that Kan Chiang's secret became common property, then suddenly we find rapid development in the technique of making iron. The stages of this, as yet not clearly defined, are associated with advances in metallurgy, the introduction of leather bellows, and improvement in the design and construction of furnices. By about 400 BC individual ironmasters were employing hundreds of men and in some of the states this industry soon became a monopoly of the government.[11] Not long after, a process for making low-grade steel was

---

[10] Quoted from Doctor Lanciotti's translation of a passage in the *Wu Yüeh Ch'un Ch'iu* in *East and West* (Year VI, no.2, pp.107–8).

[11] In his discussion of the text of the *Shih Chi* biography of Ching K'o in *Statesman, Patriot and General in Ancient China*, Professor Bodde expresses the opinion that iron was probably not widely used in China until *c.* 300 BC. Doctor Joseph Needham's monograph *Development of Iron and Steel Technology in Ancient China*, the latest scholarly survey of this subject, assigns the use of iron weapons and tools to *circa* the mid-fourth century BC. But the fact that iron tripods on which the penal laws were engraved were cast in Chin as early as 522 BC (TC, 29th year of Duke Chao) proves that the Chinese were familiar

perfected in Ch'u and Han, where 'white blades' were made. The steel-tipped lances of Ch'u were 'sharp as a bee's sting'.[12]

The effects of this break-through were soon apparent. Iron implements for agriculture and weapons of uniform high quality could be produced cheaply and in quantity, and it was natural that rulers establish foundries and arsenals and assume the prerogative (hitherto that of their vassals) of equipping the new standing armies and the conscripts. The ruler of the 'Kuo' (State) was thus enabled to assert himself more effectively than when he had not enjoyed a monopoly of weapons.

Large-scale hydraulic works, wall building, registration of the populace, and tax collection required an ever-expanding bureaucracy. Conscription and direction of the labour forces needed to carry out the grandiose schemes of the rulers, who attempted to outdo one another in the magnificence of their palaces, terraces, parks, and towers, posed complicated administrative problems. As these were solved a science of organization was created.

A parallel development took place in the military sphere in which, no less than in other proliferating areas of state activity, a high degree of administrative and directive competence was essential. By the middle of the fifth century BC large armies, which with porters and carters numbered several hundred thousand, embarked on distant campaigns which had necessarily to be planned and supported on a rational basis. When not campaigning, the army laboured (as it does in China today) on public projects.

This was an expanding society in which intellectuals, artists,

---

with crude smelting and casting processes at a much earlier date. It is scarcely conceivable that almost one hundred and fifty years would have elapsed between practical knowledge of such a process and its general application.

[12] HFT ii, p.235.

technicians, and administrators were able to employ their talents to a remarkable degree. The chronicles which record the great natural disasters, wars, assassinations of princes and ministers and usurpations occasionally remark laconically that All-under-Heaven was in chaos. This phrase is appropriate to describe the explosive age during which the last vestiges of the ancient structure were being cast aside. The hordes of experts, sophists, and disputatious scholars who offered to hard-bitten rulers and cynical ministers advice on every conceivable subject kept the intellectual world in constant ferment. Later the Legalist Han Fei Tzu was to denounce the double-faced scholars who dwelt in caves, 'pursued private studies', 'engaged in intrigues and elaborated unorthodox views'.[13] But we may safely surmise that by no means all these intellectual entrepreneurs dwelt in caves.

Confucius had wandered from one state to another in a vain attempt to persuade the rulers of his time to forsake the struggle for power and return to the enlightened path of the Sage kings. But most men of this later age realized that peregrinations devoted to the promotion of pacific and ethical objectives were a waste of time. The most pressing problems were those of practical statecraft; of internal administration and foreign policy. The crucial aspects of the latter were then the same as they have always been: to preserve and enrich the state and enhance its power and influence at the expense of enemies either actual or potential.

Consequently, while moralists may frequently have been unemployed, strategists on the whole lived comfortably—so long as their advice turned out to be good. The author of 'The Art of War' was one of these men, and even though he did not in fact find a

---

[13] HFT ii, p.235.

patron in King Ho-lü of Wu, as Ssu-ma Ch'ien says, he must somewhere have found a receptive ear. Otherwise his words would have died as did those of most of his less original contemporaries.

## IV
## WAR IN SUN TZU'S AGE

We can appreciate the originality of Sun Tzu's thought only if we are aware of the qualitative differences which distinguished warfare of the fifth and fourth centuries from that of the earlier period. Until about 500 BC war was in a sense ritualistic. Seasonal campaigns were conducted in accordance with a code generally accepted. Hostilities were prohibited during the months devoted to planting and harvesting. In winter the peasants hibernated in their mud huts; it was too cold to fight. In summer it was too hot. In theory at least, war was interdicted during the months of mourning which followed the death of a feudal lord.[1] In battle it was forbidden to strike elderly men or further injure an enemy previously wounded. The human-hearted ruler did not 'massacre cities', 'ambush armies', or 'keep the army over the season', nor did a righteous prince stoop to deceit; he did not take unfair advantage of his adversary.[2]

When King Chuang of Ch'u laid siege to the capital of Sung in 594 BC his army began to run short of provisions. To Tzu-fan, his

---

[1] This interdiction was not always observed.

[2] Dubs, i, p.167. An absurd example of this occurred in 638 BC when Duke Hsiang of Sung faced the Ch'u army at the River Hung. When the Ch'u force was half across his minister urged him to attack. The Duke refused. When the entire enemy army had crossed but was not yet arrayed for battle, the minister again pressed the Duke to attack. Duke Hsiang silenced his importunate adviser: 'The sage does not crush the feeble, nor give the order for attack until the enemy have formed their ranks.' The Duke was wounded and his forces scattered in defeat. In this context, Mao Tse-tung's often quoted remark, 'We are not the Duke of Sung', is interesting.

Minister of War, he said: 'If we exhaust these supplies without reducing the city we are going to withdraw and return home.' He then ordered Tzu-fan to climb the ramp thrown up against the wall to observe the besieged. The Prince of Sung sent his minister, Hua Yuan, to the mound to intercept him, and the following conversation ensued:

Tzu-fan said, 'How are things with your state?' Hua Yuan said, 'We are exhausted! We exchange our children and eat them, splitting and cooking the bones.'

Tzu-fan said, 'Alas! Extreme straits indeed! However, I have heard that in besieged states they gag their horses when they give them grain and send out the fat ones to meet the enemy. Now, how is it that you, Sir, are so frank?'

Hua Yuan said, 'I have heard that the superior man, seeing another's distress, has compassion on him; while the mean man, seeing another's distress, rejoices in it. I saw that you seemed to be a superior man, and that is why I was so frank.'

Tzu-fan said, 'It is so. May you exert yourself. Our army has only seven days' rations.'

Tzu-fan reported to King Chuang. King Chuang said, 'How are they?'

Tzu-fan said, 'They are exhausted. They exchange children and eat them, splitting and cooking the bones.'

King Chuang said, 'Alas! Extreme straits indeed. Now all we have to do is take them and return.'

Tzu-fan said, 'We cannot do it. I have already told them that our army for its part has only seven days' rations.'

King Chuang was angry and said, 'I sent you to observe them. Why did you tell them?'

Tzu-fan said, 'If a state as small as Sung still has a subject who does not practice deceit, how can Ch'u lack them? This is why I told him.'

King Chuang said, 'Nevertheless we shall presently just take them and return.'

Tzu-fan said, 'Let Your Highness stay here; I will just go home, if I may.'

The King said, 'If you return, leaving me, with whom shall I stay here? I shall return as you wish.' Whereupon he went back with his army.

The superior man approves their making peace themselves. Hua Yuan told Tzu-fan the truth and succeeded thereby in raising the siege and keeping intact the fortune of the two states.

Philosophers and kings distinguished between righteous and unrighteous war; an enlightened prince was morally justified in attacking 'a darkened and rustic country', in civilizing barbarians, in punishing the wilfully blind, or in dealing summarily with a state going to ruin. Such chastisements were in accord with the Will of Heaven, and were properly inflicted by the ruler in person or by a specially deputized minister of state. The commanders of the several columns were members of the hereditary aristocracy; rank in the military hierarchy directly reflected status in the feudal society. Maspero has illustrated this in an interesting study which shows how command of the Army of the Centre in Chin was from 573 BC for more than a century monopolized by a few great families.[3]

The armies of this ancient China were private in the sense that the feudal levies of Europe were private. When called upon by the sovereign, members of the nobility were supposed to provide certain numbers of chariots, horses, carts, oxen, armed footmen, grooms, cooks, and porters. The strength and nature of these contingents varied with the size of the fiefs, and as these ranged from holdings of several scores of families to several thousands, the arrays which converged upon the appointed rendezvous were

---

[3] CA, p.265 n.l.

# CH'UN CH'IU PERIOD

722–481 BC

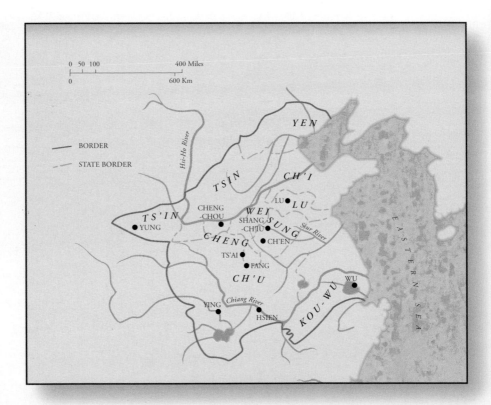

# THE CONTENDING STATES

## BOUNDARIES OF 350 BC

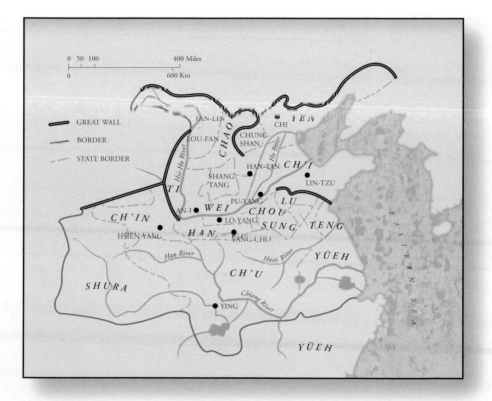

motley, to say the least. As a peasant was far less valuable than an ox or a horse, his welfare was not a matter of particular concern. The illiterate and docile serfs played but a small part in the battles of the time, in which the principal role was reserved to the four-horse chariot manned by a driver, a spearman, and a noble archer. The expendable footmen, generally protected only by padded jackets, were grouped about the chariots. A small proportion of selected men carried shields woven of bamboo or at best more cumbrous ones of crudely tanned ox or rhinoceros hide. Their arms were daggers and short swords, bronze-tipped spears, and hooking and cutting blades tied with leather thongs to wooden shafts. The bow was the weapon of the noble.

Terrain suitable for chariots dictated and restricted the form of battle; this in turn hindered the development of tactical elements. The structure of feudal society prevented the emergence of professional officers from other than noble families.

The battles of ancient China were primitive mêlées which usually produced no decisive results. Ordinarily the two sides encamped opposite one another for several days while the diviners studied the auguries and the respective commanders conducted propitiatory sacrifices. When the auspicious moment selected by the soothsayers arrived, the entire array, whose roars shook the heavens, threw itself precipitately upon the enemy. A local decision was produced speedily. Either the attacker was repulsed and allowed to withdraw, or he broke through the defender's formations, killed those still inclined to offer active resistance, pursued the flying remnants for half a mile, picked up anything of value, and returned to his own camp or capital. Victory was rarely exploited—limited operations were undertaken to achieve limited objectives.

Shortly before 500 BC the concepts which had moderated warfare began to change. War became more ferocious. A battle in 518 BC between the armies of Wu and Ch'u provides a macabre illustration. Here the Viscount of Wu ordered three thousand condemned men to line themselves up in front of his formation where in the full sight of the opposing host they committed suicide by cutting their own throats. The Ch'u army and its allied troops fled in terror.[4]

When Sun Tzu appeared on the scene, the feudal structure, in the ultimate stages of disintegration, was being replaced by an entirely different type of society in which there was much more opportunity for a talented individual. This process was gradual, but in every sphere, including the military, originality and enterprise received their rewards.

As the temporary levies of earlier days, both unreliable and inefficient, were no longer adequate, the great states formed standing armies, officered by professionals. Conscription of peasants was introduced; the new armies were composed of disciplined and well-trained troops plus conscripts whose ages ranged from sixteen to sixty. These armies were spearheaded by *élite* or shock troops specially selected for their courage, skill, discipline, and loyalty. The first such formations appeared about 500 BC and attracted enough attention for Mo Tzu to observe that King Ho-lü had trained his troops for seven years and that his *élite* corps was able to march three hundred *li* (about a hundred miles) without resting! The 'Guards' of Ch'u wore armour and helmets and carried crossbows with fifteen feathered arrows, extra arrow heads, swords and a three days' supply of parched rice. At about this time

---

[4] TC, 23rd year Duke Chao (CC V, ii, p.696). This source does not relate that the condemned criminals cut their throats, but only that they charged the allied troops fiercely. Possibly the tale was embellished by a later chronicler upon whom the commentator relied.

light troops also appeared. With standing armies incorporating these elements, operations were no longer confined to particular periods of the year. The army in being, able to take the field at short notice, was a constant threat to potential enemies.

The day of the brave, or knight, whose fame derived from his individual prowess was over. Preliminary combats between individuals, a feature common to all feudal societies, did occasionally take place. But some generals refused to put up with this.

> When Wu Ch'i fought against Ch'in, there was an officer who before battle was joined was unable to control his ardour. He advanced and took a pair of heads and returned. Wu Ch'i ordered him to be beheaded.
>
> The Army Commissioner admonished him, saying: 'This is a talented officer; you should not behead him.'
>
> Wu Ch'i replied: 'I am confident he is an officer of talent, but he is disobedient.'
>
> Thereupon he beheaded him.[5]

Battles had become directed efforts; the valiant no longer advanced unsupported, nor did the coward flee.

Elements of the new armies, capable of co-ordinated movement in accordance with detailed plans, were responsive to systematic signals. The science (or art) of tactics was born. The enemy, engaged by the *cheng* (orthodox) force, was defeated by the *ch'i* (unorthodox, unique, rare, wonderful) force, or forces; the normal pattern was a holding or fixing effort by the *cheng* while *ch'i* groups attacked the deep flanks and rear. Distraction assumed great importance and the enemy's communications became a primary target.

Although many interesting and important questions relating to

---

[5] Related by Tu Mu under v. 18, ch. vii.

details of tactics cannot be answered, we know that time and space factors were nicely calculated. Convergence of several columns upon a selected objective at a predetermined time was a technique the Chinese had mastered in Sun Tzu's day.

The concept of a 'general's staff' originated during the Warring States. Staffs included numerous specialists: weather forecasters, map makers, commissary officers, and engineers to plan tunnelling and mining operations. Others were experts on river crossing, amphibious operations, inundating, attack by fire, and the use of smoke.

Because the core of the army consisted of trained professionals and represented a considerable investment, special attention had to be paid to morale, to feeding the troops well, to rewards and punishments clearly fixed and equitably administered. Thus the spirit of the army was nurtured and at the behest of its commander it would go through fire or water. Soldiers who distinguished themselves were rewarded and promoted. This slowly but inexorably contributed to undermining the position of the hereditary aristocracy in the army.

The doctrine of collective responsibility in battle was probably first developed about this time. Commanders who retreated without orders were executed. If a section retired and its leader remained to fight, the four who had abandoned him were summarily beheaded. If a brigade or column commander withdrew without orders, he lost his head. Still, the promulgation of military codes however severe was a step forward. And while some generals enforced these more stringently than others it was recognized that arbitrary terrorism could not be relied on to produce the will to fight. Just as the professionalization of the army had opened the door to talented men, so to some extent it also

57

inhibited generals and officers from inflicting unreasonably cruel punishments or unnecessary hardships.

Naturally, all generals of the fourth century BC did not attain the position by reason of their abilities. But by this time it was possible for an able man to rise to command rank without respect to aristocratic origin and to receive at a ceremonial investiture the battle-axe which symbolized his status as commander-in-chief and conferred upon him supreme authority outside the capital. The administration of the army and its operational employment were from this moment his responsibility; when the general crossed the borders there were some orders of the ruler which he might ignore. But with his subordinates the general was amenable to the military law.

Technical improvements made their contribution to the revolution of Chinese warfare. The introduction of the crossbow and of cutting weapons of high-quality iron capable of taking and holding an edge were especially significant. Long before the crossbow appeared, the composite reflex bow had become common.[6] The crossbow, a Chinese invention of the early fourth century BC, fired heavy arrows which would have made collanders of Greek or Macedonian shields. It is likely that trained crossbowmen finally put the chariot out of business.

The armies familiar to Sun Tzu were composed of swordsmen, archers, spearmen (or halberdiers), crossbowmen, and chariots. Cavalry did not appear until some time later, but mounted men

---

[6] Reflex, or so-called 'Tartar', bows were used by the Shang (Yin) people before the Chou conquest. Whether these were then composite, i.e. of laminated horn, sinew, and wood, is not known. But later bows were constructed on this principle and to this design with pulls of well over one hundred (and sometimes as high as one hundred and fifty) pounds. Obviously these were much more powerful weapons than the single-stave bow ordinarily used in the west. The Chinese also used pellet bows, but probably principally for hunting birds.

riding without saddles or stirrups were used as scouts and messengers. Foot soldiers used two types of spear, one about 18 feet long, one about 9. These spears combined a thrusting point with a hooking or slicing blade. Spears were not used as missile weapons because in the crossbow the Chinese possessed a short-range, flat-trajectory weapon of great accuracy and tremendous striking power.

Field operations were often conducted from fortified camps designed like a Chinese city: a square enclosed by tamped earthen walls surrounded by a moat. Intersecting streets or parades, running north to south and east to west, provided for interlocking bands of fire. In the centre the commander-in-chief's banner flew over his headquarters, which was encircled by the decorated tents of his advisers and the *élite* swordsmen of his personal guard.

Before the army marched from its camp it assembled and listened to the exhortations of the general, who thundered the righteousness of the cause and excoriated the barbaric enemy. Officers feasted and exchanged vows over bloodied war drums. While troops drank wine, their ardour was aroused by gyrating sword dancers.

A Chinese army of the Warring States in battle array was an impressive spectacle as its ranks stood firm and scores of lavishly embroidered flags and banners whipped in the wind. These, decorated with figures of tigers, birds, dragons, snakes, phoenixes, and tortoises, marked the command post of the commander-in-chief behind the centre and those of the assistant generals who commanded the wings. Constant manoeuvring distracted the enemy and provided opportunities for *ch'i* operations against his deep flanks and rear.

The organization described by Sun Tzu permitted considerable flexibility in march formations, while articulation made possible rapid deployment into those suitable for battle. The five-man squad or section could obviously march either in rank or file. What was the distribution of weapons? Were crossbowmen and archers formed into separate contingents or were they organic to the section composed of a 'pair' and a 'trio'? These terms suggest they were, but from the scanty information available it appears that by the time of the battle of Ma Ling (341 BC) they were separately grouped.

What was the effective range of bows and crossbows? Here again there are no reliable data; the figures given in the records must be viewed with suspicion. It is said, for instance, that the crossbow had a range of six hundred paces. This is a gross exaggeration if lethal range is the criterion. Striking power was measured in terms of how many shields an arrow could pierce at distances of several hundred paces, but as the sort of shield used for such testing is not precisely described this fact is of little value. Nevertheless, these were powerful weapons.

That siege methods had reached a refined stage is attested by several surviving fragments of Mo Tzu's works in which various types of special apparatus for assaulting walled cities are mentioned. Scaling ladders had been in use for some centuries before his time. Movable multi-platformed towers which could be placed against city walls are referred to in The Book of Songs, as are mobile protective 'tortoises' designed to shield tunnellers.[7] We find additional information on sieges in The Book of Lord Shang. In a besieged city the entire population was mobilized and three

---

[7] CC IV, ii, iii, p.455, Ode 7.

[8] It is impossible to deal with Chinese siege and defence techniques of this period except in general terms. There is little specific information.

armies created. These consisted of able-bodied men who with abundant provisions and sharp weapons awaited the enemy; able-bodied women who erected earthworks and dug pits and moats, and children and old and feeble men and women who fed, watered, and guarded the livestock.[8]

We find in the *Sun Tzu* a doctrine relating to tactical reconnaissance, to observation, and to flank patrolling, all measures designed to ensure security on the march and in camp. Probing and testing the enemy was an essential preliminary to combat.

Thus, by the beginning of the fourth century BC, or even some decades earlier, war in China had reached a mature form; a form which indeed was not, except for employment of cavalry, to be significantly altered for many hundreds of years. At this time the Chinese possessed weapons and were masters of offensive and defensive tactics and techniques which would have enabled them to cause Alexander a great deal more trouble than did the Greeks, the Persians, or the Indians.

## V

## SUN TZU ON WAR

The opening verse of Sun Tzu's classic is the basic clue to his philosophy. War is a grave concern of the state; it must be thoroughly studied. Here is recognition—and for the first time—that armed strife is not a transitory aberration but a recurrent conscious act and therefore susceptible to rational analysis.

Sun Tzu believed that the moral strength and intellectual faculty of man were decisive in war, and that if these were properly applied war could be waged with certain success. Never to be undertaken thoughtlessly or recklessly, war was to be preceded by measures designed to make it easy to win. The master conqueror frustrated his enemy's plans and broke up his alliances. He created cleavages between sovereign and minister, superiors and inferiors, commanders and subordinates. His spies and agents were active everywhere, gathering information, sowing dissension, and nurturing subversion. The enemy was isolated and demoralized; his will to resist broken. Thus without battle his army was conquered, his cities taken and his state overthrown. Only when the enemy could not be overcome by these means was there recourse to armed force, which was to be applied so that victory was gained:

(a) in the shortest possible time;

(b) at the least possible cost in lives and effort;

(c) with infliction on the enemy of the fewest possible casualties.

National unity was deemed by Sun Tzu to be an essential requirement of victorious war. This could be attained only under a government which was devoted to the people's welfare and did not oppress them. Sun Hsing-yen was justified in observing that Sun Tzu's theories were based on 'benevolence and righteousness'.

By relating war to the immediate political context, that is to alliances or the lack of them, and to unity and stability on the home front and high morale in the army as contrasted with disunity in the enemy country and low morale in his army, Sun Tzu attempted to establish a realistic basis for a rational appraisal of relative power. His perception that mental, moral, physical, and circumstantial factors operate in war demonstrates a remarkable acuity. Few military writers, including those most esteemed in the West, have stated this proposition as clearly as did Sun Tzu some twenty-three hundred years ago. Although Sun Tzu may not have been the first to realize that armed force is the ultimate arbiter of inter-state conflicts, he was the first to put the physical clash in proper perspective.

Sun Tzu was aware of the economic implications of war. His references to inflated prices, rates of wastage, difficulties of supply, and the inevitable burdens laid upon the people show that he recognized the importance of these factors which until fairly recently have been frequently neglected.

Sun Tzu appreciated the difference between what we today define as 'national strategy' and 'military strategy'. This comes out in his discussion of the assessment of relative strengths in chapter i. Here he names five 'matters' to be deliberated in the temple councils. These are human (morale and generalship), physical (terrain and weather), and doctrinal. Only if superiority in these is clearly indicated did the council proceed to its calculations relative to numerical strengths (which Sun Tzu did not deem decisive); quality of troops; discipline, equity in the administration of rewards and punishments; and training.

Finally, this ancient writer did not conceive the object of military action to be the annihilation of the enemy's army, the destruction of his cities, and the wastage of his countryside. 'Weapons are ominous tools to be used only when there is no alternative.'

Tzu-lu, a disciple of Confucius, once discussed war with the Master:

> Tzu-lu said, Supposing you had command of the Three Hosts, whom would you take to help you? The Master said, The man who was ready to beard a tiger or rush a river without caring whether he lived or died—that sort of man I should not take. I should certainly take someone who approached difficulties with due caution and who preferred to succeed by strategy.[1]

---

[1] *Analects*, book vii, trans. Waley, p.224.

All warfare is based on deception. A skilled general must be master of the complementary arts of simulation and dissimulation; while creating shapes to confuse and delude the enemy he conceals his true dispositions and ultimate intent. When capable he feigns incapacity; when near he makes it appear that he is far away; when far away, that he is near. Moving as intangibly as a ghost in the starlight, he is obscure, inaudible. His primary target is the mind of the opposing commander; the victorious situation, a product of his creative imagination. Sun Tzu realized that an indispensable preliminary to battle was to attack the mind of the enemy.

The expert approaches his objective indirectly. By selection of a devious and distant route he may march a thousand *li* without opposition and take his enemy unaware. Such a commander prizes above all freedom of action. He abhors a static situation and therefore attacks cities only when there is no alternative. Sieges, wasteful both of lives and time, entail abdication of the initiative.

The wise general cannot be manipulated. He may withdraw, but when he does, moves so swiftly that he cannot be overtaken. His retirements are designed to entice the enemy, to unbalance him, and to create a situation favourable for a decisive counter-stroke. They are, paradoxically, offensive. He conducts a war of movement; he marches with divine swiftness; his blows fall like thunderbolts 'from the nine-layered heavens'. He creates conditions certain to produce a quick decision; for him victory is the object of war, not lengthy operations however brilliantly conducted. He knows that prolonged campaigns drain the treasury and exhaust the troops; prices rise, the people are hungry: 'No country has ever benefited from a protracted war.'

The expert commander strikes only when the situation assures victory. To create such a situation is the ultimate responsibility of generalship. Before he gives battle the superior general causes the enemy to disperse. When the enemy disperses and attempts to defend everywhere he is weak everywhere, and at the selected points many will be able to strike his few.

But vulnerability is not measured solely in physical terms. An opposing commander may be vacillating, rash, impulsive, arrogant, stubborn, or easily deceived. Possibly some elements of his army are poorly trained, disaffected, cowardly, or ineptly commanded. He may have selected a poor position. He may be overextended, his supplies low, his troops exhausted. These conditions constitute voids and provide opportunity for an imaginative general to devise an advantageous course of action.

The same factors determine the 'shape' of the opposing armies. The prudent commander bases his plans on his antagonist's 'shape'. 'Shape him', Sun Tzu says. Continuously concerned with observing and probing his opponent, the wise general at the same time takes every possible measure designed to prevent the enemy from 'shaping' him.

The actions of the general's tactical instruments—the normal, direct or *cheng* force and the extraordinary, indirect or *ch'i* force—are reciprocal; their effects are mutually reproductive. We may define the *cheng* element as fixing and the *ch'i* as flanking or encircling, or, again, as the force(s) of distraction and the force(s) of decision. Their blows are correlated. The *cheng* and the *ch'i* are compared to two interlocked rings: 'Who can tell where one begins and the other ends?' Their possible permutations are infinite; the *cheng* effort may be transformed into a *ch'i*, a *ch'i* into a *cheng*. Thus we may redefine a *ch'i* attack as one made where a decision is speedily attainable at least cost, in an area characterized by voids or fissures in the enemy's defences.

A *ch'i* operation is always unexpected, strange, or unorthodox; a *cheng,* more obvious. When Sun Tzu said to engage with the *cheng* but to win with the *ch'i* he was implying that distractive effects are necessary to ensure that decisive blows may be struck where the enemy is least prepared and where he does not anticipate them. It is misleading, however, to limit the connotation of these terms by identification with tactical battle groupings only. *Ch'i* and *cheng* operations may be launched as well on strategic levels.

Sun Tzu sees the business of a general to consist, in part, of creating changes and of manipulating them to his advantage. The excellent general weighs the situation before he moves. He does not blunder aimlessly into baited traps. He is prudent, but not hesitant. He realizes that there are 'some roads not to be followed, some armies not to be attacked, some cities not to be besieged, some positions not to be contested and some commands of the sovereign not to be obeyed'. He takes calculated risks but never needless ones. He does not 'beard a tiger or rush a river without caring whether he lives or dies'. When he sees opportunity he acts swiftly and decisively.

Sun Tzu's theory of adaptability to existing situations is an important aspect of his thought. Just as water adapts itself to the conformation of the ground, so in war one must be flexible; he must often adapt his tactics to the enemy situation. This is not in any sense a passive concept, for if the enemy is given enough rope he will frequently hang himself. Under certain conditions one yields a city, sacrifices a portion of his force, or gives up ground in order to gain a more valuable objective. Such yielding therefore masks a deeper purpose, and is but another aspect of the intellectual pliancy which distinguishes the expert in war.

Sun Tzu recognizes the hazards and the advantages of weather.

He is equally concerned with the effect of ground. The general who can assess the value of ground manoeuvres his enemy into dangerous terrain and keeps clear of it himself. He chooses the ground on which he wishes to engage, draws his enemy to it, and there gives battle. To Sun Tzu, a general unable to use ground properly was unfit to command.[2]

Sun Tzu's chapter on secret operations, as pertinent today as when he composed it, requires little elaboration except possibly to point out that he was fully aware of the necessity for compartmentation and the need for multi-level operations. Nor should his emphasis on doubled agents escape our attention. Fifth columns were as common in ancient China as in the Greek world and Sun Tzu takes account of them. The West has had considerable experience of this technique in recent years and our efforts to combat it cannot be described as entirely successful. Possibly Tu Mu's analysis of the types of men most susceptible to subversion is still worthy of examination.

With this essay, which horrified many orthodox Confucians, Sun Tzu closed his 'Art of War'.

---

[2] The Chinese have always had a special feeling for nature; for their great mountains, rivers, forests, and gorges. This is reflected in their painting, history, poetry, and other literature. Possibly the ability of their great soldiers to use terrain to best advantage derives from this apparently innate appreciation of it. China's greatest military geographer, Ku Tsu-yu (Ku Chin-fang) (1631–92), whose father and grandfather were also geographers, wrote in the Preface to his *Outline of Historical Geography*, completed *circa* 1678: Anyone who is to start military operations in one part of the country should know the condition of the country as a whole. To start such an operation without such a knowledge is to court defeat regardless of whether it is a defensive or offensive operation. (*China's Ancient Military Geography*, p.4.) Ku had great respect for Sun Tzu's appreciation of the influence terrain always has on strategy: No one can discuss strategy better than Sun Tzu and no one can discuss the advantages of terrain better than he. (Ibid., p.20.)

## VI
# SUN TZU AND MAO TSE-TUNG

Mao Tse-tung has been strongly influenced by Sun Tzu's thought. This is apparent in his works which deal with military strategy and tactics and is particularly evident in *On Guerrilla Warfare*, *On the Protracted War*, and *Strategic Problems of China's Revolutionary War*; it may also be traced in other essays less familiar to Western readers. Some years before Chairman Mao took his writing-brush in hand in Yenan, Red commanders had applied Sun Tzu's precepts to their operations in Kiangsi and Fukien, where between 1930 and 1934 they inflicted repeated defeats on Chiang K'aishek's Nationalists whose object it was to exterminate the Communists.

Mao has described his youth as punctuated by violent quarrels with an overbearing father; as a boy, he discovered an ally in his mother, whose policy of 'indirect attack'[1] appealed to him. Of his early schooling, he once remarked that it served one useful purpose—it gave him sufficient command of the basic classical literature to provide quotations as ammunition for his frequent skirmishes.

The classical canon thus found limited favour with Mao. What interested him most were 'the romances of old China and especially stories of rebellions'.[2] Among others, he read and read again the *Shui Hu Chuan* (translated by Pearl Buck as *All Men Are Brothers)* and *San Kuo* (translated by Brewitt-Taylor as *Romance of the Three Kingdoms)* and was much influenced by them'.[3] The *San Kuo* recounts battles, stratagems, and deceptions of such famous Three Kingdoms figures as Chu-ko Liang, Ts'ao Ts'ao, Lu Sun, Ssu-ma I,

---

[1] Snow, *Red Star over China*, p.128. [2] Ibid., p.130. [3] Ibid., p.131.

and Liu Pei, each of whom was a lifelong student of Sun Tzu's classic. From these stories Mao absorbed much of the military lore of his country.

During five years at the Hunan Provincial Normal School in Ch'ang-sha, Mao read in translation works of the most important Western political thinkers, but it was to the history of his country that he constantly returned. The T'aip'ing rebellion (1851–64) has always been one of Mao's favourite subjects and Li Hsiu-ch'en, the most competent leader the rebels produced, was one of his early heroes. Li, a studious man, had a remarkable flair for command. He and other T'aip'ing generals were well versed in the ancient military writers, whose precepts they turned to good account. The rebel commanders:

> always selected and advanced to the spot where the resistance was the weakest. They knew how to avoid or by-pass a strong defense and to assault a weak spot . . . They knew how to make a detour in order to attack the rear or flank of the enemy's position and how to confuse the enemy by attacking at one point to divert his attention while actually advancing on another. . . . They knew how to spy on their enemies and the activities of their fifth columnists usually preceded a formal military operation.[4]

From the T'aip'ings Mao also inherited ideas which were later reflected in his agrarian policies as well as in the rules of behaviour he incorporated into the 'Ten Commandments' of the Red Army.

Mao apparently observed that Sun Tzu's precepts are readily adaptable to the conduct of war of either the hot or cold variety, and although it was to be many years before he had the opportunity to apply them in the cold war against foreign 'imperialists', he

---

[4] Teng Ssu-yu, *New Light on the History of the T'ai P'ing Rebellion*, p.65.

had not long to wait for a chance to use them with startling effect against Chiang K'ai-shek in a hot one.

Shortly after the Nanch'ang uprising in August 1927 Mao was proscribed by the Nanking government and a price put on his head. In the early winter of that year, penniless but still confident, he arrived at the mountain base of Ching Kang Shan in the Hunan–Kiangsi borderlands. Here he was elected commander of the Red Army, which then consisted of a few thousand half-starved, miserably equipped men who had survived the Nanch'ang affair. After considerable persuasion two locally powerful bandit chieftains agreed to join forces with the Communists. These three ill-assorted groups, armed with bows and arrows, spears, antique fowling pieces, several hundred rifles, and half a dozen machine guns, formed the nucleus of the Red Army. In the spring of 1928 Chu Teh arrived at the mountain stronghold with several thousand men, most of whom were better armed, and shortly thereafter two half-hearted Nationalist attempts on the base area were repulsed.

Gradually these two men began to mould an army. Both insisted that the peasant volunteers be treated with decency and justice. Physical brutality was outlawed, as were the discriminatory practices and favouritism which chronically plagued the Manchu, Republican, and Kuomintang military establishments. Both Mao and Chu Teh (who took command of the army at this time) realized the need for a literate and well-indoctrinated force. This concern with morale, traceable in part at least to Sun Tzu's teachings, was to pay handsome dividends, for it was the major factor which preserved the Red Army after the disastrous reverses suffered in Hunan in August and early September 1930.

The birth date of the type of strategy and tactics associated with the name of Mao Tse-tung was 13 September 1930. Until that time Mao and Chu Teh had been responsive to the directives they received from the Central Committee of the Party headed by Li Li-san. These called for concentration of Red strength against cities, whose masses Li—in strict obedience to Marxist–Leninist revolutionary dogma—conceived to be the only proper base for the eventual communization of China.[5]

Here it is not necessary to describe in detail the events which led to the repudiation of the Li Li-san line by the top field command. Suffice it to say that August and September 1930 were the most critical months in the history of the Chinese Communist Party. After assaulting Ch'angsha and holding the city for a few days, the Reds under P'eng Te-huai were driven out. Two hundred miles to the east, at Nanch'ang, repeated assaults of the principal force under Mao and the army commander, Chu Teh, were bloodily repulsed.

The unimaginative stubbornness of Li Li-san almost succeeded in breaking the back of the Red Army. But not quite. Before this happened Mao and Chu Teh made the decision to break off the Nanch'ang action and withdraw. The Li Li-san group, however, insisted that the attack on Ch'angsha should be resumed, and much against their better judgement Mao and Chu Teh accepted these orders. The result of a week's fighting was almost catastrophic, and in the evening of 13 September 1930 the Red Army, shattered by a month of almost continuous battle against greatly superior forces, withdrew toward central Kiangsi. In October Chiang launched the first of the so-called 'extermination campaigns'. A new phase now began.

While the Communist veterans of the campaigns in south China have a right to be proud of their successes in the mobile war they

---

[5] Li Li-san was following the Moscow directive.

conducted, these must be evaluated in the light of the opposition the Red Army was required to face during the first four 'Bandit Suppression' campaigns directed with such ineptitude from Nanking.

The majority of the non-Central divisions of the 1930's, and particularly the 'war lord' troops, were composed of illiterate peasant conscripts. These men, poorly trained and equipped, badly fed and irregularly paid, were treated in arbitrary fashion by their officers. The rate of desertion was fantastic, and as padded muster rolls were normal it was impossible to determine the actual strength of any Kuomintang unit. Normally complemented at about ten thousand, many divisions could at best have paraded half that number. Peculation was endemic; it afflicted most of the officers including many of the generals. Nepotism was rampant. In many units the venereal disease rate was shockingly high. Medical facilities were almost totally lacking. It is not surprising that the morale of such troops left something to be desired. Their feeble officers did little to improve the situation.

At this time there were half a dozen 'model' Central Army divisions commanded by competent, brave, and honest generals. Not until the Fourth Campaign was any of these committed.[6] Fortunately for the Communists, they rarely encountered Kuomintang divisions of this standard. By committing poor units under inept commanders to his 'suppression' campaigns, the Generalissimo contributed to the steady increase in Red strength.

---

[6] The first campaigns were fought by a Red Army greatly inferior in every material respect. This army had no aircraft, no motor vehicles, no tactical radios or telephones, no artillery, no medical service, few mortars, a limited number of light and heavy machine guns, and was always plagued with a critical shortage of ammunition. It survived not entirely because of the poor quality of the opposition, but in some measure because of the intellectual flexibility of its commanders, the morale of its rank and file, and the superior intelligence which resulted from the support of the peasants. It enjoyed as well a decided superiority in tactical doctrine.

In this policy there was more than meets the eye: Chiang's idea was that the Reds and the non-Central troops would destroy each other. But the troops did not see the situation in precisely the same way. They surrendered to the Communists by battalions. Many of the captured officers and men immediately joined the Red Army. The weapons taken were numbered in tens of thousands. In 1936 Mao remarked:

> We have a claim on the output of the arsenals of London as well as of
>
> Hanyang, and, what is more, it is to be delivered to us by the enemy's
>
> own transport corps. This is the sober truth, not a joke.[7]

By 1949 the Americans, who had spent several billion dollars equipping, training, supporting, and transporting Chiang's armies, were fully aware that this was indeed no joke.

Only during the Fifth Campaign (planned by his German advisers and mounted in late 1933) was the Generalissimo able to impose his will on the Reds. Nationalist troops, including well-trained and equipped Central divisions, were used in overwhelming force; their advances were slow, careful, and co-ordinated. As they inched methodically south they applied a scorched-earth policy. Peasants were forcibly removed from the zone of operations; the Communists were deprived of their ability to get information. The Nationalists had finally learned how essential it was to maintain contact between adjacent elements; the Reds were thus not able to concentrate against isolated units and overwhelm them. For the first time they suddenly discovered that they had been deprived of the initiative. The result was unexpected: the Communists panicked and cracked. Reduced to complete passivity, they did not, in Mao's words, 'show the slightest initiative or

---

[7] Mao Tse-tung, *Selected Works*, i, p.253.

dynamic force', and were left with 'no alternative but to withdraw' from Kiangsi. This campaign, later the subject of countless exhaustive post mortems, forced the Red command to undertake the now celebrated Long March to north-west China.

As their later operations proved, the Reds had learned; one lesson they carried with them into Shensi was that those deprived of the initiative usually lose, those who retain it usually win. Loss of initiative in the Fifth Campaign was in part due to overconfidence; the Red high command committed the cardinal sin of underestimating the enemy. Here, for the first time, the Reds knew neither the enemy nor themselves, and were in peril in every battle. Possibly with this experience in mind Mao later wrote:

> We must not belittle the saying in the book of Sun Wu Tzu, the great
> military expert of ancient China, 'Know your enemy and know your-
> self and you can fight a hundred battles without disaster.'[8]

Kiangsi and the 'Long March' were the Communists' military laboratories; Yenan, the quiet retreat in which experience was analysed. Shortly after arriving at Pao An, the temporary Red capital, Mao took to his cave and began to write. He devoted little time to analysis of successes; study of the failures was more rewarding. With disarming honesty he described the last of the 'Bandit Suppression' campaigns as a Red 'fiasco' in which the Communists had neglected to observe the principle which should govern all military operations:

> The first essential of military operations is to preserve one's own forces
> and annihilate the enemy and to attain this end it is necessary to . . .
> avoid all passive and inflexible methods. . . .[9]

The question of the offence as opposed to passive defence did

---

[8] Mao Tse-tung, *Selected Works*, i, p.187.  [9] Mao Tse-tung, *Selected Works*, ii, p.96.

not worry Mao; he realized, as Sun Tzu had, that no war can be won by adoption of a static attitude. On this subject the Chairman minced no words; he describes those who deliberately assume such a position as 'fools'.

The strategy and tactics used with such success against the Japanese emphasized constant movement and were based on four slogans coined at Ching Kang Shan:

1. When the enemy advances, we retreat!
2. When the enemy halts, we harass!
3. When the enemy seeks to avoid battle, we attack!
4. When the enemy retreats, we pursue!

Mao has never felt it necessary to point out the remarkable similarity of his sixteen-character jingle to several of Sun Tzu's verses.

Later, when Mao was able to reflect fully on the lessons of the battles in the south and on the Long March he wrote, in paraphrased elaboration of Sun Tzu:

In general, the shifting of forces should be done secretly and swiftly. Ingenious devices such as making a noise in the east while attacking in the west, appearing now in the south and now in the north, hit-and-run and night action should be constantly employed to mislead, entice and confuse the enemy.

Flexibility in dispersion, in concentration and in shifting is the concrete manifestation of the initiative in guerrilla warfare, whereas inflexibility and sluggishness will inevitably land one in a passive position and incur unnecessary losses. But a commander proves himself wise not by understanding how important the flexible employment of forces is but by being able to disperse, concentrate or shift his forces in time according to specific circumstances. This wisdom in foreseeing changes and right timing is not easy to acquire except for those who study with a receptive mind and take pains to investigate and think things over. In order that flexibility may not become reckless action, a careful consideration of the circumstances is necessary.[10]

Communist commanders repeatedly proved themselves capable of utilizing terrain more effectively than their opponents. Real estate, as such, was never an important factor with the Reds, who were experts at running away. Mao has several times humorously remarked that he doubted very much if any army had ever been quite so proficient in this respect. But this running away was usually designed to draw the enemy on and to induce over-confidence in his commanders, who became arrogant and lax. Sucked into unknown country, deprived of information, and with tenuous lines of communication, Nationalist units in Kiangsi were skilfully 'cut out' and dealt with individually. This process was to be applied with equal success during the Civil War, both in Manchuria and North China.

The superior intelligence service of the Communists usually enabled them to determine the enemy's 'shape'; their own they were equally successful in obscuring. Their appraisals of the enemy were almost invariably accurate. Mao later wrote:

> Some people are intelligent in knowing themselves but stupid in know-
> ing their opponents, and others the other way round; neither kind can
> solve the problem of learning and applying the laws of war.[11]

One of the most difficult problems which confronts any commander who has committed his forces in accordance with a well-developed plan is to alter this in the light of changing circumstances. Sun Tzu recognized the inherent difficulties, both intellectual and physical, and repeatedly emphasized that the nature of war is ceaseless change. For this reason operations

---

[10] Mao Tse-tung, *Selected Works*, ii, pp.130–1.

[11] Mao Tse-tung, *Selected Works*, i, p.187.

require continuous review and readjustment. Mao writes:

> The process of knowing the situation goes on not only before but also
> after the formulation of a military plan. The carrying out of a plan,
> from its very beginning to the conclusion of an operation, is another
> process of knowing the situation, i.e., the process of putting it into
> practice. In this process, there is need to examine anew whether the
> plan mapped out in the earlier process corresponds with the actualities.
> If the plan does not correspond or does not fully correspond with
> them, then we must, according to fresh knowledge, form new judge-
> ments and make new decisions to modify the original plan in order to
> meet the new situation. There are partial modifications in almost every
> operation, and sometimes even a complete change. A hothead who
> does not know how to change his plan, or is unwilling to change it but
> acts blindly, will inevitably run his head against a brick wall.[12]

This seems unnecessarily verbose, but history provides ample
evidence that the theme needs to be repeated again and again.

To retire when conditions indicate it to be desirable is correct;
attack and defence are complementary. Mao paraphrases Sun Tzu
this way:

> Attack may be changed into defense and defense into attack; advance
> may be turned into retreat and retreat into advance; containing forces
> may be turned into assault forces, and assault forces into containing
> forces.[13]

It is one of the most important tasks of command 'to effect
timely and proper change of tactics according to the conditions of
the units and of the terrain, both on the enemy's side and our
own'.[14] One yields when it is expedient; he gives A in order that he

---

[12] Mao Tse-tung, *Selected Works*, pp.185–6. [13] Mao Tse-tung, *On the Protracted War*, pp.102–3.
[14] Ibid., pp.102–3.

may take B. By timely retirement he conserves his strength and preserves the initiative. Conversely, a belated retirement is essentially a passive action: initiative has been lost.

Deception and surprise are two key principles. Again paraphrasing Sun Tzu, Mao has said that war demands deception. 'It is often possible by adopting all kinds of measures of deception to drive the enemy into the plight of making erroneous judgements and taking erroneous actions, thus depriving him of his superiority and initiative.'[15] The enemy is deceived by creating 'shapes' (Sun Tzu) or 'illusions' (Mao).

At the same time, one conceals his shape from the enemy. The eyes and ears of hostile commanders are sealed. Deception is not enough—the enemy's leaders must be confused; if possible, driven insane.[16] The morale of the enemy is the target of high priority, its reduction an essential preliminary to the armed clash. Here again is a distinct echo of Sun Tzu, the first proponent of psychological warfare.

From Mao's work, man emerges as the decisive factor in war. Weapons are important but not decisive. It is man's directing intelligence which counts most:

> In actual life we cannot ask for an invincible general; there have been few such generals since ancient times. We ask for a general who is both brave and wise, who usually wins battles in the course of a war— a general who combines wisdom with courage.[17]

The wise general is circumspect; he prefers to succeed by strategy:

> We do not allow any of our Red Army commanders to become rash and reckless hot-heads and must encourage everyone of them to

---

<sup>15</sup> Mao Tse-tung, *On the Protracted War*, p.98.  <sup>16</sup> Ibid., p.100.
<sup>17</sup> Mao Tse-tung, *Selected Works*, i, p.183.

become a hero who, at once brave and wise, possesses not only the courage to override all obstacles but the ability to control the changes and developments of an entire war.[18]

This ability is what Sun Tzu had in mind when he used the phrase 'to control victory'.

The dispositions of a thoughtful commander 'ensue from correct decisions' derived from 'correct judgements', which depend on 'a comprehensive and indispensable reconnaissance'.[19] The data gathered by observation and from reports are carefully appraised; the crude and false discarded; the refined and true retained. The wise general thus is able to go 'through the outside into the inside'.[20] A careless one 'bases his military plan upon his own wishful thinking'; it does not correspond with reality; it is, in a word, 'fantastic'.[21]

In the early phases of the Civil War the Reds repeatedly demonstrated their mastery of deceptive tactics and their mobility. An almost uncanny ability to determine points of Nationalist weakness permitted them to exploit these qualities and led inevitably to an accelerating disintegration of the Nationalist position. Throughout the Civil War the Communists continually threw Sun Tzu's book of war at the Generalissimo's dispirited commanders. In Manchuria, to which he permitted himself to be lured despite the advice of his American advisers, the 'Gimo's' best divisions were macerated and his hopes destroyed.

In Korea the Chinese Communists deployed almost a quarter of a million men to battle positions south of the Yalu before the United Nations command became aware that its widely dispersed

---

[18] Mao Tse-tung, *Selected Works*, i, p.188.

[19] Mao Tse-tung, *Selected Works*, i, p.185.

[20] Ibid., p.185. [21] Ibid., p.185.

elements were even seriously threatened. This grand manoeuvre, imaginatively conceived and skilfully executed, was the preliminary to a driving offensive that came within an ace of destroying the United Nations force in Korea.

But except for the short period which immediately followed intervention, the People's Liberation Army was forced to fight under stabilized conditions. Circumstances did not permit the PLA to conduct the mobile war for which it was best fitted. Faced with constricted terrain which favoured a technically skilled opponent capable of deploying massive resources of fire power, the Red command had little scope for ingenuity. The adoption of wave tactics seems in retrospect to have been almost an act of desperation.

Some Western observers have drawn conclusions from the latter phases of the Korean experience which would not necessarily be applicable in other situations. It is dangerous to assume that the Chinese will operate in accordance with any previous pattern. It is safer to expect them to change their tactics 'in an infinite number of ways'. Mao has said:

> We should carefully study the lessons which were learned in past wars
> at the cost of blood and which have been bequeathed to us. . . .We
> must put conclusions thus reached to the test of our own experiences

---

[22] Mao Tse-tung, *On the Protracted War*, p.186.

[23] One of those most responsible for the great interest the Chinese Communists have displayed in Sun Tzu is Kuo Hua-jo, whose name is practically unknown in the West. In 1939 Kuo completed an analytical commentary on 'The Thirteen Chapters' entitled *A Preliminary Study of Sun Tzu's Art of War* (*Sun Tzu Ping Fa Ch'u Pu Yen Chiu*) (孫子兵法初步研究). This was designed to be used as a military textbook in Red-controlled areas. The position Kuo has now enjoyed as a leading military theoretician seems to date from that period. His latest edition of 'The Art of War' is *Chin I Hsin P'en Sun Tzu Ping Fa* (今譯新編孫子兵法). *A Modern Translation with New Chapter Arrangement of Sun Tzu's 'Art of War'*. As the title suggests, the material has been completely rearranged. Sun Tzu's verses have been phrased in colloquial Chinese and simplified characters are used throughout.

and absorb what is useful, reject what is useless and add what is specifically our own.[22]

It has often been said that had Western leaders read Hitler's *Mein Kampf* they would have been somewhat better equipped than they were to deal with him. Some familiarity with Mao's speeches and writings, together with the major works which provide their conceptual framework, would assist leaders of the present generation to an equal degree. From any collection of such works, 'The Art of War' could not be omitted.[23]

傳記

# BIOGRAPHY OF
# SUN TZU[1]

Sun Tzu was a native of Ch'i who by means of his book on the art of war secured an audience with Ho-lü, King of Wu.[2]

Ho-lü said, 'I have read your thirteen chapters, Sir, in their entirety.[3] Can you conduct a minor experiment in control of the movement of troops?'

Sun Tzu replied, 'I can.'

Ho-lu asked, 'Can you conduct this test using women?'

Sun Tzu said, 'Yes.'

The King thereupon agreed and sent from the palace one hundred and eighty beautiful women.[4]

Sun Tzu divided them into two companies and put the King's two favourite concubines in command. He instructed them all how to hold halberds. He then said, 'Do you know where the heart is, and where the right and left hands and the back are?'

The women said, 'We know.'

Sun Tzu said, 'When I give the order "Front", face in the direction of the heart, when I say "Left", face toward the left hand; when I say "Right" toward the right; when I say "Rear", face in the direction of your backs.'

The women said, 'We understand.'

---

[1] As it appears in the SC, *Sun Tzu Wu Ch'i Lieh Chuan* (孫子吳起列傳).

[2] A normal procedure to secure a powerful patron.

[3] This statement proves only that there were thirteen chapters at the time Ssu-ma Ch'ien wrote the SC.

[4] According to another version, there were three hundred.

When these regulations had been announced the executioner's weapons were arranged.[5]

Sun Tzu then gave the orders three times and explained them five times, after which he beat on the drum the signal 'Face Right'. The women all roared with laughter.

Sun Tzu said, 'If regulations are not clear and orders not thoroughly explained, it is the commander's fault.' He then repeated the orders three times and explained them five times, and gave the drum signal to face to the left. The women again burst into laughter.

Sun Tzu said, 'If instructions are not clear and commands not explicit, it is the commander's fault. But when they have been made clear, and are not carried out in accordance with military law, it is a crime on the part of the officers.' Then he ordered that the commanders of the right and left ranks be beheaded.

The King of Wu, who was reviewing the proceedings from his terrace, saw that his two beloved concubines were about to be executed. He was terrified, and hurriedly sent an aide with this message: 'I already know that the General is able to employ troops. Without these two concubines my food will not taste sweet. It is my desire that they be not executed.'

Sun Tzu replied: 'Your servant has already received your appointment as Commander and when the commander is at the head of the army he need not accept all the sovereign's orders.'

Consequently he ordered that the two women who had commanded the ranks be executed as an example. He then used the next seniors as company commanders.

Thereupon he repeated the signals on the drum, and the women faced left, right, to the front, to the rear, knelt and rose all

---

[5] To make it clear that he meant business.

84

in strict accordance with the prescribed drill. They did not dare to make the slightest noise.

Sun Tzu then sent a messenger to the King and informed him: 'The troops are now in good order. The King may descend to review and inspect them. They may be employed as the King desires, even to the extent of going through fire and water.'

The King of Wu said, 'The General may go to his hostel and rest. I do not wish to come to inspect them.'

Sun Tzu said, 'The King likes only empty words. He is not capable of putting them into practice.'

Ho-lü then realized Sun Tzu's capacity as a commander, and eventually made him a general. Sun Tzu defeated the strong State of Ch'u to the west and entered Ying; to the north he intimidated Ch'i and Chin.[6] That the name of Wu was illustrious among the feudal lords was partly due to his achievements. (The Yüeh *Chüeh Shu* says: 'Outside the Wu Gate of Wu Hsieh, at a distance of ten *li*, there was a large tomb which is that of Sun Tzu.')[7]

Something more than one hundred years after Sun Tzu's death, there was a Sun Pin, who was born between O and Chuan.[8] Sun Pin was a descendant of Sun Tzu. He and P'ang Chüan studied military theory together. P'ang Chüan served in the State of Wei and obtained the position of commander from King Hui.[9] He realized that his ability was not equal to that of Sun Pin, and secretly sent a messenger to summon him. When Pin arrived, P'ang Chüan, fearing that he was worthier than himself, was jealous and

---

[6] No other historical record substantiates these statements.

[7] The sentence in brackets was added as a textual note by Sun Hsing-yen. The work he cited is probably a forgery of the fourth century BC or later. Wu Hsieh is modern Soochow.

[8] In ancient Wei. Chuan was near P'o Hsien in modern Shantung.

[9] Became King of Wei in 344 BC.

trumped up a charge against him, so that Sun Pin's feet were cut off and his face branded, and he was hidden away where he could not be seen.[10]

An ambassador from Ch'i came to Ta Liang.[11] Sun Pin, who was in the status of a criminal, saw him secretly and talked to him. The Ch'i ambassador thought him an extraordinary person and carried him off to Ch'i by stealth.[12]

T'ien Chi, the commander-in-chief of Ch'i, was pleased with him and treated him as a guest.

T'ien Chi frequently gambled on horse races with the Princes of Ch'i. Sun Pin noticed that the teams of horses did not differ greatly. The horses were of three classes, first, second, and third. Observing this, Sun Pin said to T'ien Chi, 'You place a bet on this contest; your servant can make you win.'

T'ien Chi believed him and agreed with the King and the Princes on a wager of one thousand pieces of gold on the races. As he was about to put up his money, Sun Pin said, 'Match your third string against his first, your best against his second, and your second best against his weakest.' The three competitions were completed, and while T'ien Chi did not win the first, he won the last two, and a thousand pieces of gold from the King. Thereupon

---

[10] *Pin* (臏) means 'kneecap'. But it also means 'to cut off the feet'. This was the third of the five mutilating punishments, which in order of severity were: (a) Tattooing, or branding, the face. (b) Cutting off the nose. (c) Cutting off the feet. (d) Castration (or, for women, claustration). (e) Death.

Sun 'The Footless' would not have been so called until after his mutilation. A criminal sentenced to be castrated sometimes also suffered the lesser punishments of branding, cutting off the nose, and severance of the feet.

[11] Capital of the Wei State. Its site was near modern K'aifeng in Honan Province. States were often designated by the names of their capitals, as they are today.

[12] This statement does not agree with other sources in which we are told that after suffering the punishment described, as the result of royal censure initiated by P'ang Chüan's slanders, he feigned insanity and escaped to Ch'i, where he was put in charge of military affairs.

T'ien Chi introduced Sun Pin to King Wei, who discussed military matters with him and made him a staff officer.

Later, when the State of Wei attacked Chao, Chao urgently appealed to Ch'i for help.[13] King Wei desired to make Sun Pin the commander-in-chief. Sun Pin declined with thanks and said: 'Since I have previously been a convict this would not be appropriate.' T'ien Chi was then made commander-in-chief, and Sun Pin was appointed his chief of staff.

Sun Pin travelled in a baggage wagon, and made his plans while sitting. T'ien Chi wished to lead the army into Chao State. Sun Pin said: 'One who wishes to unravel the confused and entangled does not grasp the entire skein. Similarly, to dissolve a combat, one does not grasp the halberds. Strike at a salient or at an unprotected place. Then, when the antagonists have reached a stalemate, the situation will resolve itself. Now Wei and Chao attack one another. The light formations and shock troops are in the field and are certainly exhausted; at home the old and weak are tired. Nothing is better than to march speedily on Ta Liang, seizing the principal routes and roads, and attack the capital while it is unprotected. Then Wei must disengage from Chao in order to save itself. Thus in one stroke we may lift the siege of Chao and at the same time reap the fruit of Wei's defeat.'

T'ien Chi followed this advice with the result that the army of Wei left Han Tai and engaged Ch'i at Kuei Ling, where the Wei troops were severely defeated.

Fifteen years later, Wei in alliance with Chao attacked Han.[14] Han urgently appealed to Ch'i for aid. The King of Ch'i ordered T'ien Chi to take the field and march directly on Ta Liang.

---

[13] In 356 BC. [14] In 341 BC.

The Wei commander-in-chief, P'ang Chüan, hearing of this, left Han and returned to his own country. The army of Ch'i had already crossed the border and was marching to the west.

Sun Pin spoke to T'ien Chi and said: 'The troops of the three Chin States are usually fierce, brave and contemptuous of Ch'i.[15] They consider Ch'i to be cowardly. The skilful fighter will take this circumstance into account and plan his strategy to profit from it. According to 'The Art of War', if the army presses on to gain advantage from a distance of one hundred *li,* the commander of the van will be captured;[16] if from fifty *li,* only half the troops will reach the critical point.' He then ordered that when the Ch'i army crossed the borders and entered Wei, they should on the first night build one hundred thousand kitchen fires, on the following night fifty thousand, and on the third, thirty thousand.

P'ang Chüan marched for three days and, greatly pleased, said: 'I have always been certain that the troops of Ch'i were cowards. They have been in my country for only three days and more than half their officers and soldiers have deserted!'

He thereupon left behind his heavy infantry and wagons, and with lightly armed shock troops only followed by forced marches.[17] Sun Pin had calculated that P'ang Chüan would arrive at Ma Ling in the evening. The Ma Ling road is narrow, and on both sides there are many gorges and defiles where troops may be placed in ambush.

Sun Pin cut the bark off a great tree, and on the trunk wrote: 'P'ang Chüan dies under this tree.' He then placed the most

---

[15] Han, Wei, and Chao are always referred to as 'The Three Chins'.

[16] Lit. 'the commander of the Van will be torn away'.

[17] The SC reads 'infantry wagons [chariots]' or 'infantry and wagons [chariots]'. I assume this means that heavily armed and armoured infantry and the supply trains were left while P'ang Chüan pressed on.

skilful archers of the army with ten thousand crossbows in ambush on both sides of the road, and ordered that when in the evening they saw fire, all were to shoot at it. P'ang Chüan actually arrived that night and when he saw writing on a tree, ignited a torch to read what was written there. Before he had finished the ten thousand crossbowmen of Ch'i discharged their arrows simultaneously, and the army of Wei was thrown into the utmost confusion. P'ang Chüan, at his wit's end, realized that his troops would be defeated. Whereupon he cut his throat, and as he expired, said: 'So I have contributed to the fame of that wretch!' Sun Pin, taking advantage of this victory, completely destroyed the Wei army and captured the heir apparent, Shen, after which he returned to Ch'i.[18]

Because of this, Sun Pin's reputation was world wide and generations have transmitted his strategy.

---

[18] This great battle took place in north-east Shantung in 341 BC.

存神過化豈容巖穴列千

# ESTIMATES[1]

Sun Tzu said:

## 1

War is a matter of vital importance to the State; the province of life or death; the road to survival or ruin.[2] It is mandatory that it be thoroughly studied.

*Li Ch'üan: 'Weapons are tools of ill omen.' War is a grave matter; one is apprehensive lest men embark upon it without due reflection.*

## 2

Therefore, appraise it in terms of the five fundamental factors and make comparisons of the seven elements later named.[3] So you may assess its essentials.

## 3

The first of these factors is moral influence; the second, weather; the third, terrain; the fourth, command; and the fifth, doctrine.[4]

---

[1] The title means 'reckoning', 'plans', or 'calculations'. In the Seven Military Classics edition the title is 'Preliminary Calculations'. The subject first discussed is the process we define as an Estimate (or Appreciation) of the Situation.

[2] Or 'for [the field of battle] is the place of life and death [and war] the road to survival or ruin'.

[3] Sun Hsing-yen follows the *T'ung T'ien* here and drops the character *shih* (事): 'matters', 'factors', or 'affairs'. Without it the verse does not make much sense.

[4] Here *Tao* (道) is translated 'moral influence'. It is usually rendered as 'The Way', or 'The Right Way'. Here it refers to the morality of government; specifically to that of

*Chang Yü: The systematic order above is perfectly clear. When troops are raised to chastise transgressors, the temple council first considers the adequacy of the rulers' benevolence and the confidence of their peoples; next, the appropriateness of nature's seasons, and finally the difficulties of the topography. After thorough deliberation of these three matters a general is appointed to launch the attack.⁵ After troops have crossed the borders, responsibility for laws and orders devolves upon the general.*

## 4

By moral influence I mean that which causes the people to be in harmony with their leaders, so that they will accompany them in life and unto death without fear of mortal peril.⁶

*Chang Yü: When one treats people with benevolence, justice, and righteousness, and reposes confidence in them, the army will be united in mind and all will be happy to serve their leaders. The Book of Changes says: 'In happiness at overcoming difficulties, people forget the danger of death.'*

---

the sovereign. If the sovereign governs justly, benevolently, and righteously, he follows the Right Path or the Right Way, and thus exerts a superior degree of moral influence. The character *fa* (法), here rendered 'doctrine', has as a primary meaning 'law' or 'method'. In the title of the work it is translated 'Art'. But in v. 8 Sun Tzu makes it clear that here he is talking about what we call doctrine.

⁵ There are precise terms in Chinese which cannot be uniformly rendered by our word 'attack'. Chang Yu here uses a phrase which literally means 'to chastise criminals', an expression applied to attack of rebels. Other characters have such precise meanings as 'to attack by stealth', 'to attack suddenly', 'to suppress the rebellious', 'to reduce to submission', &c.

⁶ Or 'Moral influence is that which causes the people to be in accord with their superiors. . . .' Ts'ao Ts'ao says the people are guided in the right way (of conduct) by 'instructing' them.

## 5

By weather I mean the interaction of natural forces;
the effects of winter's cold and summer's heat and
the conduct of military operations in accordance with
the seasons.[7]

## 6

By terrain I mean distances, whether the ground is tra-
versed with ease or difficulty, whether it is open or
constricted, and the chances of life or death.

*Mei Yao-ch'en: . . . When employing troops it is
essential to know beforehand the conditions of the
terrain. Knowing the distances, one can make use of
an indirect or a direct plan. If he knows the degree
of ease or difficulty of traversing the ground he can
estimate the advantages of using infantry or cavalry. If he knows where the ground is constricted and
where open he can calculate the size of force appropriate. If he knows where he will give battle he
knows when to concentrate or divide his forces.*[8]

## 7

By command I mean the general's qualities of wisdom,
sincerity, humanity, courage, and strictness.

*Li Ch'üan: These five are the virtues of the general.
Hence the army refers to him as 'The Respected One'.
Tu Mu : . . . If wise, a commander is able to recog-
nize changing circumstances and to act expediently.
If sincere, his men will have no doubt of the cer-
tainty of rewards and punishments. If humane, he*

---

[7] It is clear that the character *t'ien* (天) (Heaven) is used in this verse in the sense of 'weather', as it is today.

[8] Knowing the ground of life and death. . .' is here rendered 'If he knows where he will give battle'.

*loves mankind, sympathizes with others, and appreciates their industry and toil. If courageous, he gains victory by seizing opportunity without hesitation. If strict, his troops are disciplined because they are in awe of him and are afraid of punishment.*

*Shen Pao-hsu. . . said: 'If a general is not courageous he will be unable to conquer doubts or to create great plans.'*

## 8

By doctrine I mean organization, control, assignment of appropriate ranks to officers, regulation of supply routes, and the provision of principal items used by the army.

## 9

There is no general who has not heard of these five matters. Those who master them win; those who do not are defeated.

## 10

Therefore in laying plans compare the following elements, appraising them with the utmost care.

## 11

If you say which ruler possesses moral influence, which commander is the more able, which army obtains the advantages of nature and the terrain, in which regulations and instructions are better carried out, which troops are the stronger;[9]

**Chang Yü:** *Chariots strong, horses fast, troops valiant, weapons sharp—so that when they hear the drums beat the attack they are happy, and when they hear the gongs sound the retirement they are enraged. He who is like this is strong.*

## 12

Which has the better trained officers and men;

**Tu Yu:** *. . . Therefore Master Wang said: 'If officers are unaccustomed to rigorous drilling they will be worried and hesitant in battle; if generals are not thoroughly trained they will inwardly quail when they face the enemy.'*

---

[9] In this and the following two verses the seven elements referred to in v. 2 are named.

## 13

And which administers rewards and punishments in a
more enlightened manner;
***Tu Mu:*** *Neither should be excessive.*

## 14

I will be able to forecast which side will be victorious
and which defeated.

## 15

If a general who heeds my strategy is employed he is
certain to win. Retain him! When one who refuses to
listen to my strategy is employed, he is certain to be
defeated. Dismiss him!

## 16

Having paid heed to the advantages of my plans, the
general must create situations which will contribute to
their accomplishment.[10] By 'situations' I mean that he
should act expediently in accordance with what is
advantageous and so control the balance.

## 17

All warfare is based on deception.

## 18

Therefore, when capable, feign incapacity; when active,
inactivity.

## 19

When near, make it appear that you are far away; when
far away, that you are near.

---

[10] Emending *i* (以) to *i* (已). The commentators do not agree on an interpretation
of this verse.

# 20

Offer the enemy a bait to lure him; feign disorder and strike him.

*Tu Mu: The Chao general Li Mu released herds of cattle with their shepherds; when the Hsiung Nu had advanced a short distance he feigned a retirement, leaving behind several thousand men as if abandoning them. When the Khan heard this news he was delighted, and at the head of a strong force marched to the place. Li Mu put most of his troops into formations on the right and left wings, made a horning attack, crushed the Huns and slaughtered over one hundred thousand of their horsemen.[11]*

# 21

When he concentrates, prepare against him; where he is strong, avoid him.

# 22

Anger his general and confuse him.

*Li Ch'üan: If the general is choleric his authority can easily be upset. His character is not firm.*

*Chang Yü: If the enemy general is obstinate and prone to anger, insult and enrage him, so that he will be irritated and confused, and without a plan will recklessly advance against you.*

# 23

Pretend inferiority and encourage his arrogance.

*Tu Mu: Toward the end of the Ch'in dynasty, Mo Tun of the Hsiung Nu first established his power. The Eastern Hu were strong and sent ambassadors*

---

[11] The Hsiung Nu were nomads who caused the Chinese trouble for centuries. The Great Wall was constructed to protect China from their incursions.

*to parley. They said: 'We wish to obtain T'ou Ma's thousand-li horse.' Mo Tun consulted his advisers, who all exclaimed: 'The thousand-li horse! The most precious thing in this country! Do not give them that!' Mo Tun replied: 'Why begrudge a horse to a neighbour?' So he sent the horse.*[12]

*Shortly after, the Eastern Hu sent envoys who said: 'We wish one of the Khan's princesses.' Mo Tun asked advice of his ministers who all angrily said: 'The Eastern Hu are unrighteous! Now they even ask for a princess! We implore you to attack them!' Mo Tun said: 'How can one begrudge his neighbour a young woman?' So he gave the woman.*

*A short time later, the Eastern Hu returned and said: 'You have a thousand li of unused land which we want.' Mo Tun consulted his advisers. Some said it would be reasonable to cede the land, others that it would not. Mo Tun was enraged and said: 'Land is the foundation of the State. How could one give it away?' All those who had advised doing so were beheaded. Mo Tun then sprang on his horse, ordered that all who remained behind were to be beheaded, and made a surprise attack on the Eastern Hu. The Eastern Hu were contemptuous of him and had made no preparations. When he attacked he annihilated them. Mo Tun then turned westward and attacked the Yueh Ti. To the south he annexed Lou Fan . . . and invaded Yen. He completely recovered the ancestral lands of the Hsiung Nu previously conquered by the Ch'in general Meng T'ien.*[13]

**Ch'en Hao:** *Give the enemy young boys and*

---

[12] Mo Tun, or T'ou Ma or T'ouman, was the first leader to unite the Hsiung Nu. The thousand-*li* horse was a stallion reputedly able to travel a thousand *li* (about three hundred miles) without grass or water. The term indicates a horse of exceptional quality, undoubtedly reserved for breeding.

*women to infatuate him, and jades and silks to excite his ambitions.*

## 24

Keep him under a strain and wear him down.

**Li Ch'üan:** *When the enemy is at ease, tire him.*

**Tu Mu:** *. . . Toward the end of the Later Han, after Ts'ao Ts'ao had defeated Liu Pei, Pei fled to Yuan Shao, who then led out his troops intending to engage Ts'ao Ts'ao. T'ien Fang, one of Yuan Shao's staff officers, said: 'Ts'ao Ts'ao is expert at employing troops; one cannot go against him heedlessly. Nothing is better than to protract things and keep him at a distance. You, General, should fortify along the mountains and rivers and hold the four prefectures. Externally, make alliances with powerful leaders; internally, pursue an agro-military policy.[14] Later, select crack troops and form them into extraordinary units. Taking advantage of spots where he is unprepared, make repeated sorties and disturb the country south of the river. When he comes to aid the right, attack his left; when he goes to succour the left, attack the right; exhaust him by causing him continually to run about. . . . Now if you reject this victorious strategy and decide instead to risk all on one battle, it will be too late for regrets.' Yuan Shao did not follow this advice and therefore was defeated.[15]*

---

[13] Meng T'ien subdued the border nomads during the Ch'in, and began the construction of the Great Wall. It is said that he invented the writing-brush. This is probably not correct, but he may have improved the existing brush in some way.

[14] This refers to agricultural military colonies in remote areas in which soldiers and their families were settled. A portion of the time was spent cultivating the land, the remainder in drilling, training, and fighting when necessary. The Russians used this policy in colonizing Siberia. And it is in effect now in Chinese borderlands.

[15] During the period known as 'The Three Kingdoms', Wei in the north and west, Shu in the south-west, and Wu in the Yangtze valley contested for empire.

## 25

When he is united, divide him.

*Chang Yü: Sometimes drive a wedge between a sovereign and his ministers; on other occasions separate his allies from him. Make them mutually suspicious so that they drift apart. Then you can plot against them.*

## 26

Attack where he is unprepared;
sally out when he does not expect you.

*Ho Yen-hsi: . . . Li Ching of the T'ang proposed ten plans to be used against Hsiao Hsieh, and the entire responsibility of commanding the armies was entrusted to him. In the eighth month he collected his forces at K'uei Chou.[16]*

*As it was the season of the autumn floods the waters of the Yangtze were overflowing and the roads by the three gorges were perilous, Hsiao Hsieh thought it certain that Li Ching would not advance against him. Consequently he made no preparations.*

*In the ninth month Li Ching took command of the troops and addressed them as follows: 'What is of the greatest importance in war is extraordinary speed; one cannot afford to neglect opportunity. Now we are concentrated and Hsiao Hsieh does not yet know of it. Taking advantage of the fact that the river is in flood, we will appear unexpectedly under the walls of his capital. As is said: 'When the thunder-clap comes, there is no time to cover the ears.' Even if he should discover us, he cannot on the spur of the moment devise a plan to counter us, and surely we can capture him.'*

---

[16] K'uei Chou is in Ssu Ch'uan.

*He advanced to I Ling and Hsiao Hsieh began to be afraid and summoned reinforcements from south of the river, but these were unable to arrive in time. Li Ching laid siege to the city and Hsieh surrendered.*

*'To sally forth where he does not expect you' means as when, toward its close, the Wei dynasty sent Generals Chung Hui and Teng Ai to attack Shu.[17] . . . In winter, in the tenth month, Ai left Yin P'ing and marched through uninhabited country for over seven hundred* li, *chiselling roads through the mountains and building suspension bridges. The mountains were high, the valleys deep, and this task was extremely difficult and dangerous. Also, the army, about to run out of provisions, was on the verge of perishing. Teng Ai wrapped himself in felt carpets and rolled down the steep mountain slopes; generals and officers clambered up by grasping limbs of trees. Scaling the precipices like strings of fish, the army advanced.*

*Teng Ai appeared first at Chiang Yu in Shu, and Ma Mou, the general charged with its defence, surrendered. Teng Ai beheaded Chu-ko Chan, who resisted at Mien-chu, and marched on Ch'eng Tu. The King of Shu, Liu Shan, surrendered.*

---

[17] This campaign was conducted *c.* AD 255.

[18] A confusing verse difficult to render into English. In the preliminary calculations some sort of counting devices were used. The operative character represents such a device, possibly a primitive abacus. We do not know how the various 'factors' and 'elements' named were weighted, but obviously the process of comparison of relative strengths was a rational one. It appears also that two separate calculations were made, the first on a national level, the second on a strategic level. In the former the five basic elements named in v. 3 were compared; we may suppose that if the results of this were favourable the military experts compared strengths, training, equity in administering rewards and punishments, and so on (the seven factors).

# 27

These are the strategist's keys to victory. It is not
possible to discuss them beforehand.

*Mei Yao-ch'en: When confronted by the enemy
respond to changing circumstances and devise expedients. How can these be discussed beforehand?*

# 28

Now if the estimates made in the temple before
hostilities indicate victory it is because calculations
show one's strength to be superior to that of his enemy;
if they indicate defeat, it is because calculations show
that one is inferior. With many calculations, one
can win; with few one cannot. How much less chance
of victory has one who makes none at all! By this
means I examine the situation and the outcome will
be clearly apparent.[18]

# WAGING WAR

Sun Tzu said:

## 1

Generally, operations of war require one thousand fast four-horse chariots, one thousand four-horse wagons covered in leather, and one hundred thousand mailed troops.

*Tu Mu:* . . . *In ancient chariot fighting, 'leather-covered chariots' were both light and heavy. The latter were used for carrying halberds, weapons, military equipment, valuables, and uniforms. The Ssu-ma Fa said: 'One chariot carries three mailed officers; seventy-two foot troops accompany it. Additionally, there are ten cooks and servants, five men to take care of uniforms, five grooms in charge of fodder, and five men to collect firewood and draw water. Seventy-five men to one light chariot, twenty-five to one baggage wagon, so that taking the two together one hundred men compose a company.[1]*

## 2

When provisions are transported for a thousand *li* expenditures at home and in the field, stipends for the entertainment of advisers and visitors, the cost of materials such as glue and lacquer, and of chariots and

---

[1] The ratio of combat to administrative troops was thus 3:1.

armour, will amount to one thousand pieces of gold a day. After this money is in hand, one hundred thousand troops may be raised.[2]

**Li Ch'üan:** *Now when the army marches abroad, the treasury will be emptied at home.*

**Tu Mu:** *In the army there is a ritual of friendly visits from vassal lords. That is why Sun Tzu mentions 'advisers and visitors'.*

## 3

Victory is the main object in war.[3] If this is long delayed, weapons are blunted and morale depressed. When troops attack cities, their strength will be exhausted.

## 4

When the army engages in protracted campaigns the resources of the state will not suffice.

**Chang Yü:** . . . *The campaigns of the Emperor Wu of the Han dragged on with no result and after the treasury was emptied he issued a mournful edict.*

## 5

When your weapons are dulled and ardour damped, your strength exhausted and treasure spent, neighbouring rulers will take advantage of your distress to act. And even though you have wise counsellors, none will be able to lay good plans for the future.

---

[2] Gold money was coined in Ch'u as early as 400 BC, but actually Sun Tzu does not use the term 'gold'. He uses a term which meant 'metallic currency'.

[3] I insert the character *kuei* (貴) following the 'Seven Martial Classics'. In the context the character has the sense of 'what is valued' or 'what is prized'.

## 6

Thus, while we have heard of blundering swiftness
in war, we have not yet seen a clever operation that
was prolonged.

*Tu Yu: An attack may lack ingenuity, but it must
be delivered with supernatural speed.*

## 7

For there has never been a protracted war from which a
country has benefited.

*Li Ch'üan: The Spring and Autumn Annals says:
'War is like unto fire; those who will not put aside
weapons are themselves consumed by them.'*

## 8

Thus those unable to understand the dangers inherent
in employing troops are equally unable to understand
the advantageous ways of doing so.

## 9

Those adept in waging war do not require a
second levy of conscripts nor more than one
provisioning.[4]

## 10

They carry equipment from the homeland; they rely
for provisions on the enemy. Thus the army is plenti-
fully provided with food.

---

[4] The commentators indulge in lengthy discussions as to the number of provisionings. This version reads 'they do not require three'. That is, they require only two, i.e. one when they depart and the second when they return. In the meanwhile they live on the enemy. The TPYL version (following Ts'ao Ts'ao) reads: 'they do not require to be *again* provisioned', that is during a campaign. I adopt this.

# 11

When a country is impoverished by military operations
it is due to distant transportation; carriage of supplies
for great distances renders the people destitute.
*Chang Yü:. . . If the army had to be supplied with
grain over a distance of one thousand* li, *the troops
would have a hungry look.*[5]

---

[5] This comment appears under V. 10 but seems more appropriate here.

## 12

Where the army is, prices are high; when prices rise
the wealth of the people is exhausted. When wealth
is exhausted the peasantry will be afflicted with
urgent exactions.[6]

*Chia Lin:. . . Where troops are gathered the price of
every commodity goes up because everyone covets the
extraordinary profits to be made.[7]*

## 13

With strength thus depleted and wealth consumed the
households in the central plains will be utterly impoverished and seven-tenths of their wealth dissipated.

*Li Ch'üan: If war drags on without cessation men
and women will resent not being able to marry, and
will be distressed by the burdens of transportation.*

## 14

As to government expenditures, those due to broken-
down chariots, worn-out horses, armour and helmets,
arrows and crossbows, lances, hand and body shields,
draft animals and supply wagons will amount to sixty
per cent of the total.[8]

## 15

Hence the wise general sees to it that his troops feed on
the enemy, for one bushel of the enemy's provisions is
equivalent to twenty of his; one hundredweight of
enemy fodder to twenty hundredweight of his.

---

[6] Or, 'close to [where] the army [is]', (i.e. in the zone of operations) 'buying
is expensive; when buying is expensive. . .'. The 'urgent [or 'heavy'] exactions' refers
to special taxes, forced contributions of animals and grain, and porterage.

[7] This comment, which appears under the previous verse, has been transposed.

[8] Here Sun Tzu uses the specific character for 'crossbow'.

*Chang Yü:. . . In transporting provisions for a distance of one thousand* li, *twenty bushels will be consumed in delivering one to the army. . . . If difficult terrain must be crossed even more is required.*

## 16

The reason troops slay the enemy is because they are enraged.[9]

*Ho Yen-hsi: When the Yen army surrounded Chi Mo in Ch'i, they cut off the noses of all the Ch'i prisoners.[10] The men of Ch'i were enraged and conducted a desperate defence. T'ien Tan sent a secret agent to say: 'We are terrified that you people of Yen will exhume the bodies of our ancestors from their graves. How this will freeze our hearts!'*

*The Yen army immediately began despoiling the tombs and burning the corpses. The defenders of Chi Mo witnessed this from the city walls and with tears flowing wished to go forth to give battle, for rage had multiplied their strength by ten. T'ien Tan knew then that his troops were ready, and inflicted a ruinous defeat on Yen.*

## 17

They take booty from the enemy because they desire wealth.

*Tu Mu: . . . In the Later Han, Tu Hsiang, Prefect of Chin Chou, attacked the Kuei Chou rebels Pu Yang, P'an Hung, and others. He entered Nan Hai, destroyed three of their camps, and captured much treasure. However, P'an Hung and his followers*

---

[9] This seems out of place.

[10] This siege took place in 279 BC.

were still strong and numerous, while Tu Hsiang's troops, now rich and arrogant, no longer had the slightest desire to fight.

Hsiang said: 'Pu Yang and P'an Hung have been rebels for ten years. Both are well-versed in attack and defence. What we should really do is unite the strength of all the prefectures and then attack them. For the present the troops shall be encouraged to go hunting.' Whereupon the troops both high and low went together to snare game.

As soon as they had left, Tu Hsiang secretly sent people to burn down their barracks. The treasures they had accumulated were completely destroyed. When the hunters returned there was not one who did not weep.

Tu Hsiang said: 'The wealth and goods of Pu Yang and those with him are sufficient to enrich several generations. You gentlemen did not do your best. What you have lost is but a small bit of what is there. Why worry about it?'

When the troops heard this, they were all enraged and wished to fight. Tu Hsiang ordered the horses fed and everyone to eat in his bed, and early in the morning they marched on the rebels' camp.[11] Yang and Hung had not made preparations, and Tu Hsiang's troops made a spirited attack and destroyed them.

**Chang Yü:** . . . In this Imperial Dynasty, when the Eminent Founder ordered his generals to attack Shu, he decreed: 'In all the cities and prefectures taken, you should, in my name, empty

---

[11] They ate a pre-cooked meal in order to avoid building fires to prepare breakfast?

*the treasuries and public storehouses to entertain
the officers and troops. What the State wants is only
the land.'*

## 18

Therefore, when in chariot fighting more than
ten chariots are captured, reward those who take
the first. Replace the enemy's flags and banners with
your own, mix the captured chariots with yours,
and mount them.

## 19

Treat the captives well, and care for them.

*Chang Yü: All the soldiers taken must be cared for with magnanimity and sincerity so that they may be used by us.*

## 20

This is called 'winning a battle and becoming stronger'.

## 21

Hence what is essential in war is victory, not prolonged operations. And therefore the general who understands war is the Minister of the people's fate and arbiter of the nation's destiny.

*Ho Yen-hsi: The difficulties in the appointment of a commander are the same today as they were in ancient times.[12]*

[12] Ho Yen-hsi probably wrote this about AD 1050.

# OFFENSIVE STRATEGY

Sun Tzu said:

## 1

Generally in war the best policy is to take a state intact;
to ruin it is inferior to this.
*Li Ch'üan: Do not put a premium on killing.*

## 2

To capture the enemy's army is better than to destroy
it; to take intact a battalion, a company or a five-man
squad is better than to destroy them.

## 3

For to win one hundred victories in one hundred
battles is not the acme of skill. To subdue the enemy
without fighting is the acme of skill.

## 4

Thus, what is of supreme importance in war is to
attack the enemy's strategy;[1]
*Tu Mu: . . . The Grand Duke said: 'He who excels
at resolving difficulties does so before they arise. He
who excels in conquering his enemies triumphs
before threats materialize.'*
*Li Ch'üan: Attack plans at their inception. In the*

---

[1] Not, as Giles translates, 'to balk the enemy's plans'.

*Later Han, K'ou Hsün surrounded Kao Chun.*[2]
*Chun sent his Planning Officer, Huang-fu Wen, to*
*parley. Huang-fu Wen was stubborn and rude and*
*K'ou Hsün beheaded him, and informed Kao*
*Chun: 'Your staff officer was without propriety. I*
*have beheaded him. If you wish to submit, do so*
*immediately. Otherwise defend yourself.' On the*
*same day, Chun threw open his fortifications and*
*surrendered.*

*All K'ou Hsün's generals said: 'May we ask, you*
*killed his envoy, but yet forced him to surrender his*
*city. How is this?'*

*K'ou Hsün said: 'Huang-fu Wen was Kao Chun's*
*heart and guts, his intimate counsellor. If I had*
*spared Huang-fu Wen's life, he would have accom-*
*plished his schemes, but when I killed him, Kao*
*Chun lost his guts. It is said: "The supreme excel-*
*lence in war is to attack the enemy's plans."'*

*All the generals said: 'This is beyond our*
*comprehension.'*

## 5

Next best is to disrupt his alliances:[3]
**Tu Yu:** Do not allow your enemies to get together.
**Wang Hsi:** . . . Look into the matter of his alliances
and cause them to be severed and dissolved. If an
enemy has alliances, the problem is grave and the
enemy's position strong; if he has no alliances the
problem is minor and the enemy's position weak.

## 6

The next best is to attack his army.

---

[2] This took place during the first century AD.

[3] Not, as Giles translates, 'to prevent the junction of the enemy's forces'.

*Chia Lin:* . . . *The Grand Duke said: 'He who struggles for victory with naked blades is not a good general.'*

*Wang Hsi: Battles are dangerous affairs.*

*Chang Yü: If you cannot nip his plans in the bud, or disrupt his alliances when they are about to be consummated, sharpen your weapons to gain the victory.*

## 7

The worst policy is to attack cities. Attack cities only when there is no alternative.[4]

## 8

To prepare the shielded wagons and make ready the necessary arms and equipment requires at least three months; to pile up earthen ramps against the walls an additional three months will be needed.

## 9

If the general is unable to control his impatience and orders his troops to swarm up the wall like ants, one-third of them will be killed without taking the city. Such is the calamity of these attacks.

*Tu Mu:* . . . *In the later Wei, the Emperor T'ai Wu led one hundred thousand troops to attack the Sung general Tsang Chih at Yu T'ai. The Emperor first asked Tsang Chih for some wine.[5] Tsang Chih sealed up a pot full of urine and sent it to him. T'ai Wu was transported with rage and immediately*

---

[4] In this series of verses Sun Tzu is not discussing the art of generalship as Giles apparently thought. These are objectives or policies—*cheng* (政)—in order of relative merit.

[5] Exchange of gifts and compliments was a normal preliminary to battle.

*attacked the city, ordering his troops to scale the
walls and engage in close combat. Corpses piled up
to the top of the walls and after thirty days of this
the dead exceeded half his force.*

# 10

Thus, those skilled in war subdue the enemy's
army without battle. They capture his cities without
assaulting them and overthrow his state without
protracted operations.

*Li Ch'üan: They conquer by strategy. In the Later
Han the Marquis of Tsan, Tsang Kung, surrounded
the 'Yao' rebels at Yüan Wu, but during a succession
of months was unable to take the city.*[6] *His officers
and men were ill and covered with ulcers. The King
of Tung Hai spoke to Tsang Kung, saying: 'Now you
have massed troops and encircled the enemy, who is
determined to fight to the death. This is no strategy!
You should lift the siege. Let them know that an
escape route is open and they will flee and disperse.
Then any village constable will be able to capture
them!' Tsang Kung followed this advice and took
Yüan Wu.*

# 11

Your aim must be to take All-under-Heaven intact.
Thus your troops are not worn out and your gains will
be complete. This is the art of offensive strategy.

# 12

Consequently, the art of using troops is this: When ten
to the enemy's one, surround him;

---

[6] Yao (妖) connotes the supernatural. The Boxers, who believed themselves
impervious to foreign lead, could be so described.

# 13

When five times his strength, attack him;
*Chang Yü: If my force is five times that of the enemy
I alarm him to the front, surprise him to the rear,
create an uproar in the east and strike in the west.*

# 14

If double his strength, divide him.[7]
*Tu Yu: . . . If a two-to-one superiority is insufficient
to manipulate the situation, we use a distracting
force to divide his army. Therefore the Grand Duke
said: 'If one is unable to influence the enemy to
divide his forces, he cannot discuss unusual tactics.'*

# 15

If equally matched you may engage him.
*Ho Yen-hsi: . . . In these circumstances only the able
general can win.*

# 16

If weaker numerically, be capable of withdrawing;
*Tu Mu: If your troops do not equal his, temporari-
ly avoid his initial onrush. Probably later you can
take advantage of a soft spot. Then rouse yourself
and seek victory with determined spirit.*
*Chang Yü: If the enemy is strong and I am weak,
I temporarily withdraw and do not engage.[8] This
is the case when the abilities and courage of the
generals and the efficiency of troops are equal.*

---

[7] Some commentators think this verse 'means to divide one's own force', but that
seems a less satisfactory interpretation, as the character chih (之) used in the two
previous verses refers to the enemy.
[8] Tu Mu and Chang Yü both counsel 'temporary' withdrawal, thus emphasizing the
point that offensive action is to be resumed when circumstances are propitious.

*If I am in good order and the enemy in disarray,
if I am energetic and he careless, then, even if he be
numerically stronger, I can give battle.*

## 17

And if in all respects unequal, be capable of
eluding him, for a small force is but booty for one
more powerful.[9]

*Chang Yü: . . . Mencius said: 'The small certainly
cannot equal the large, nor can the weak match the
strong, nor the few the many.'[10]*

## 18

Now the general is the protector of the state. If this
protection is all-embracing, the state will surely be
strong; if defective, the state will certainly be weak.

*Chang Yü: . . . The Grand Duke said: 'A sovereign
who obtains the right person prospers. One who fails
to do so will be ruined.'*

## 19

Now there are three ways in which a ruler can bring
misfortune upon his army:[11]

## 20

When ignorant that the army should not advance,
to order an advance or ignorant that it should not
retire, to order a retirement. This is described as
'hobbling the army'.

強
壯
的

---

[9] Lit. 'the strength of a small force is. . .'. This apparently refers to its weapons
and equipment.  [10] CC II (Mencius), i, ch. 7.

[11] Here I have transposed the characters meaning 'ruler' and 'army', otherwise the
verse would read that there are three ways in which an army can bring misfortune
upon the sovereign.

**Chia Lin:** *The advance and retirement of the army can be controlled by the general in accordance with prevailing circumstances. No evil is greater than commands of the sovereign from the court.*

# 21

When ignorant of military affairs, to participate in their administration. This causes the officers to be perplexed.

**Ts'ao Ts'ao:** *. . . An army cannot be run according to rules of etiquette.*

**Tu Mu:** *As far as propriety, laws, and decrees are concerned, the army has its own code, which it ordinarily follows. If these are made identical with those used in governing a state the officers will be bewildered.*

**Chang Yü:** *Benevolence and righteousness may be used to govern a state but cannot be used to administer an army. Expediency and flexibility are used in administering an army, but cannot be used in governing a state.*

# 22

When ignorant of command problems to share in the exercise of responsibilities. This engenders doubts in the minds of the officers.[12]

**Wang Hsi:** *. . . If one ignorant of military matters is sent to participate in the administration of the army, then in every movement there will be disagreement and mutual frustration and the entire army will be hamstrung. That is why Pei Tu memorialized the throne to withdraw the Army Supervisor; only then was he able to pacify Ts'ao Chou.*[13]

---

[12] Lit. 'Not knowing [or 'not understanding' or 'ignorant of'] [where] authority [lies] in the army'; or 'ignorant of [matters relating to exercise of] military authority . . .'. The operative character is 'authority' or 'power'.

[13] The 'Army Supervisors' of the T'ang were in fact political commissars. Pei Tu

*Chang Yü: In recent times court officials have been used as Supervisors of the Army and this is precisely what is wrong.*

## 23

If the army is confused and suspicious, neighbouring rulers will cause trouble. This is what is meant by the saying: 'A confused army leads to another's victory.'[14]

*Meng:. . . The Grand Duke said: 'One who is confused in purpose cannot respond to his enemy.'*

*Li Ch'üan:. . . The wrong person cannot be appointed to command. . . . Lin Hsiang-ju, the Prime Minister of Chao, said: 'Chao Kua is merely able to read his father's books, and is as yet ignorant of correlating changing circumstances. Now Your Majesty, on account of his name, makes him the commander-in-chief. This is like glueing the pegs of a lute and then trying to tune it.'*

## 24

Now there are five circumstances in which victory may be predicted:

## 25

He who knows when he can fight and when he cannot will be victorious.

## 26

He who understands how to use both large and small forces will be victorious.

---

became Prime Minister in AD 815 and in 817 requested the throne to recall the supervisor assigned him, who must have been interfering in army operations.

[14] 'Feudal Lords' is rendered 'neighbouring rulers'. The commentators agree that a confused army robs itself of victory.

*Tu Yu: There are circumstances in war when many cannot attack few, and others when the weak can master the strong. One able to manipulate such circumstances will be victorious.*

## 27

He whose ranks are united in purpose will be victorious.

*Tu Yu: Therefore Mencius said: 'The appropriate season is not as important as the advantages of the ground; these are not as important as harmonious human relations.'[15]*

## 28

He who is prudent and lies in wait for an enemy who is not, will be victorious.

*Ch'en Hao: Create an invincible army and await the enemy's moment of vulnerability.*

*Ho Yen-hsi: . . . A gentleman said: 'To rely on rustics and not prepare is the greatest of crimes; to be prepared beforehand for any contingency is the greatest of virtues.'*

## 29

He whose generals are able and not interfered with by the sovereign will be victorious.

*Tu Yu:. . . Therefore Master Wang said: 'To make appointments is the province of the sovereign; to decide on battle, that of the general.'*

*Wang Hsi:. . . A sovereign of high character and intelligence must be able to know the right man, should place the responsibility on him, and expect results.*

*Ho Yen-hsi: . . . Now in war there may be one hundred changes in each step. When one sees he can, he advances; when he sees that things are difficult, he retires. To say that a general must await commands of the sovereign in such circumstances is like inform-ing a superior that you wish to put out a fire. Before the order to do so arrives the ashes are cold. And it is said one must consult the Army Supervisor in these matters! This is as if in building a house beside the road one took advice from those who pass by. Of course the work would never be completed![16] To put a rein on an able general while at the same time asking him to suppress a cunning enemy is like tying up the Black Hound of Han and then ordering him to catch elusive hares. What is the difference?*

# 30

It is in these five matters that the way to victory is known.

# 31

Therefore I say: 'Know the enemy and know yourself; In a hundred battles you will never be in peril.

# 32

When you are ignorant of the enemy but know yourself, your chances of winning or losing are equal.

# 33

If ignorant both of your enemy and of yourself, you are certain in every battle to be in peril.'
**Li Ch'üan:** *Such people are called 'mad bandits'. What can they expect if not defeat?*

---

[15] CC II (Mencius), ii, ch. 1, p. 85.

[16] A paraphrase of an ode which Legge renders:

They are like one taking counsel with wayfarers about building a house

Which consequently will never come to completion. (CC IV, ii, p. 332, Ode I.)

# DISPOSITIONS[1]

Sun Tzu said:

## 1

Anciently the skilful warriors first made themselves invincible and awaited the enemy's moment of vulnerability.

## 2

Invincibility depends on one's self; the enemy's vulnerability on him.

## 3

It follows that those skilled in war can make themselves invincible but cannot cause an enemy to be certainly vulnerable.

*Mei Yao-ch'en: That which depends on me, I can do; that which depends on the enemy cannot be certain.*

## 4

Therefore it is said that one may know how to win, but cannot necessarily do so.

---

[1] The character *hsing* (形) means 'shape', 'form', or 'appearance' or in a more restricted sense, 'disposition' or 'formation'. The Martial Classics edition apparently followed Ts'ao Ts'ao and titled the chapter *Chun Hsing* (軍形), 'Shape [or 'Dispositions'] of the Army'. As will appear, the character connotes more than mere physical dispositions.

## 5

Invincibility lies in the defence; the possibility of
victory in the attack.[2]

## 6

One defends when his strength is inadequate;
he attacks when it is abundant.

## 7

The experts in defence conceal themselves as under
the ninefold earth; those skilled in attack move as
from above the ninefold heavens. Thus they are
capable both of protecting themselves and of
gaining a complete victory.[3]

*Tu Yu: Those expert at preparing defences consider
it fundamental to rely on the strength of such obstacles as mountains, rivers and foothills. They make it
impossible for the enemy to know where to attack.
They secretly conceal themselves as under the nine-
layered ground.*

*Those expert in attack consider it fundamental
to rely on the seasons and the advantages of
the ground; they use inundations and fire according
to the situation. They make it impossible for an
enemy to know where to prepare. They release
the attack like a lightning bolt from above the nine-
layered heavens.*

## 8

To foresee a victory which the ordinary man can
foresee is not the acme of skill;

---

[2] 'Invincibility is [means] defence; the ability to conquer is [means] attack.'

[3] The concept that Heaven and Earth each consist of 'layers' or 'stages'
is an ancient one.

*Li Ch'üan:. . . When Han Hsin destroyed Chao State he marched out of the Well Gorge before breakfast. He said: 'We will destroy the Chao army and then meet for a meal.' The generals were despondent and pretended to agree. Han Hsin drew up his army with the river to its rear. The Chao troops climbed upon their breastworks and, observing this, roared with laughter and taunted him: 'The General of Han does not know how to use troops!' Han Hsin then proceeded to defeat the Chao army and after breakfasting beheaded Lord Ch'eng An.*

*This is an example of what the multitude does not comprehend.*[4]

## 9

To triumph in battle and be universally acclaimed 'Expert' is not the acme of skill, for to lift an autumn down requires no great strength; to distinguish between the sun and moon is no test of vision; to hear the thunderclap is no indication of acute hearing.[5]

*Chang Yü: By 'autumn down' Sun Tzu means rabbits' down, which on the coming of autumn is extremely light.*

## 10

Anciently those called skilled in war conquered an enemy easily conquered.[6]

---

[4] Han Hsin placed his army in 'death ground'. He burned his boats and smashed his cooking pots. The river was at the rear, the Chao army to the front. Han Hsin had to conquer or drown.

[5] To win a hard-fought battle or to win one by luck is no mark of skill.

[6] The enemy was conquered easily because the experts previously had created appropriate conditions.

# 11

And therefore the victories won by a master of
war gain him neither reputation for wisdom
nor merit for valour.

*Tu Mu: A victory gained before the situation has
crystallized is one the common man does not comprehend. Thus its author gains no reputation for
sagacity. Before he has bloodied his blade the enemy
state has already submitted.*

*Ho Yen-hsi: . . . When you subdue your enemy
without fighting who will pronounce you valorous?*

# 12

For he wins his victories without erring. 'Without
erring' means that whatever he does insures his victory;
he conquers an enemy already defeated.

*Chen Hao: In planning, never a useless move; in
strategy, no step taken in vain.*

# 13

Therefore the skilful commander takes up a position
in which he cannot be defeated and misses no
opportunity to master his enemy.

# 14

Thus a victorious army wins its victories before
seeking battle; an army destined to defeat
fights in the hope of winning.

*Tu Mu:. . . Duke Li Ching of Wei said: 'Now, the
supreme requirements of generalship are a clear
perception, the harmony of his host, a profound
strategy coupled with far-reaching plans, an understanding of the seasons and an ability to examine
the human factors. For a general unable to estimate*

*his capabilities or comprehend the arts of expedien-
cy and flexibility when faced with the opportunity
to engage the enemy will advance in a stumbling
and hesitant manner, looking anxiously first to his
right and then to his left, and be unable to produce
a plan. Credulous, he will place confidence in unreliable reports, believing at one moment this and at
another that. As timorous as a fox in advancing or
retiring, his groups will be scattered about. What is
the difference between this and driving innocent
people into boiling water or fire? Is this not exactly
like driving cows and sheep to feed wolves or tigers?'*

## 15

Those skilled in war cultivate the *Tao* and preserve
the laws and are therefore able to formulate
victorious policies.

**Tu Mu:** *The* Tao *is the way of humanity and
justice; 'laws' are regulations and institutions.
Those who excel in war first cultivate their own
humanity and justice and maintain their laws and
institutions. By these means they make their governments invincible.*

## 16

Now the elements of the art of war are first,
measurement of space; second, estimation of
quantities; third, calculations; fourth, comparisons;
and fifth, chances of victory.

## 17

Measurements of space are derived from the ground.

## 18

Quantities derive from measurement, figures from quantities, comparisons from figures, and victory from comparisons.

*Ho Yen-hsi:*[7] *'Ground' includes both distances and type of terrain; 'measurement' is calculation. Before the army is dispatched, calculations are made respecting the degree of difficulty of the enemy's land; the directness and deviousness of its roads; the number of his troops; the quantity of his war equipment and the state of his morale. Calculations are made to see if the enemy can be attacked and only after this is the population mobilized and troops raised.*

## 19

Thus a victorious army is as a hundredweight balanced against a grain; a defeated army as a grain balanced against a hundredweight.

## 20

It is because of disposition that a victorious general is able to make his people fight with the effect of pent-up waters which, suddenly released, plunge into a bottomless abyss.

*Chang Yü:* *The nature of water is that it avoids heights and hastens to the lowlands. When a dam is broken, the water cascades with irresistible force. Now the shape of an army resembles water. Take advantage of the enemy's unpreparedness; attack him when he does not expect it; avoid his strength and strike his emptiness, and like water, none can oppose you.*

---

[7] This comment appears in the text after V. 18. The factors enumerated are qualities of 'shape'.

# ENERGY[1]

Sun Tzu said:

## 1

Generally, management of many is the same as management of few. It is a matter of organization.[2]

*Chang Yü: To manage a host one must first assign responsibilities to the generals and their assistants, and establish the strengths of ranks and files. . . .*

*One man is a single; two, a pair; three, a trio. A pair and a trio make a five,[3] which is a squad; two squads make a section; five sections, a platoon; two platoons, a company; two companies, a battalion; two battalions, a regiment; two regiments, a group; two groups, a brigade; two brigades, an army.[4] Each is subordinate to the superior and controls the inferior. Each is properly trained. Thus one may manage a host of a million men just as he would a few.*

---

[1] *Shih* (勢) the title of this chapter, means 'force', influence', 'authority', 'energy'. The commentators take it to mean 'energy' or 'potential' in some contexts and 'situation' in others.

[2] *Fen Shu* (分 數) is literally 'division of [or by] numbers' (or 'division and numbering'). Here translated 'organization'.

[3] Suggestive that the 'pair' and the 'trio' carried different weapons.

[4] A ten-man section; one hundred to the company; two hundred to the battalion; four hundred to the regiment; eight hundred to the group; sixteen hundred to the brigade; thirty-two hundred to the army. This apparently reflects organization at the time Chang Yu was writing. The English terms for the units are arbitrary.

## 2

And to control many is the same as to control few.
This is a matter of formations and signals.

*Chang Yü: . . . Now when masses of troops are employed, certainly they are widely separated, and ears are not able to hear acutely nor eyes to see clearly. Therefore officers and men are ordered to advance or retreat by observing the flags and banners and to move or stop by signals of bells and drums. Thus the valiant shall not advance alone, nor shall the coward flee.*

## 3

That the army is certain to sustain the enemy's attack without suffering defeat is due to operations of the extraordinary and the normal forces.[5]

*Li Ch'üan: The force which confronts the enemy is the normal; that which goes to his flanks the extraordinary. No commander of an army can wrest the advantage from the enemy without extraordinary forces.*

*Ho Yen-hsi: I make the enemy conceive my normal force to be the extraordinary and my extraordinary to be the normal. Moreover, the normal may become the extraordinary and vice versa.*

## 4

Troops thrown against the enemy as a grindstone against eggs is an example of a solid acting upon a void.

---

[5] The concept expressed by *cheng* (正) 'normal' (or 'direct') and *ch'i* (奇), 'extraordinary' (or 'indirect') is of basic importance. The normal (*cheng*) force fixes or distracts the enemy; the extraordinary (*ch'i*) forces act when and where their blows are not anticipated. Should the enemy perceive and respond to a *ch'i* manoeuvre in such a manner as to neutralize it, the manicure would automatically become *cheng*.

*Ts'ao Ts'ao: Use the most solid to attack the most empty.*

## 5

Generally, in battle, use the normal force to engage;
use the extraordinary to win.

## 6

Now the resources of those skilled in the use of extraordinary forces are as infinite as the heavens and earth;
as inexhaustible as the flow of the great rivers.[6]

## 7

For they end and recommence; cyclical, as are the
movements of the sun and moon. They die away and
are reborn; recurrent, as are the passing seasons.

## 8

The musical notes are only five in number but
their melodies are so numerous that one
cannot hear them all.

## 9

The primary colours are only five in number but their
combinations are so infinite that one cannot
visualize them all.

## 10

The flavours are only five in number but their blends
are so various that one cannot taste them all.

---

[6] Sun Tzu uses the characters *chiang* (江) and *ho* (河) which I have rendered 'the
great rivers'.

## 11

In battle there are only the normal and extraordinary
forces, but their combinations are limitless; none can
comprehend them all.

## 12

For these two forces are mutually reproductive; their
interaction as endless as that of interlocked rings. Who
can determine where one ends and the other begins?

## 13

When torrential water tosses boulders,
it is because of its momentum;

# 14

When the strike of a hawk breaks the body of its prey,
it is because of timing.[7]

*Tu Yu: Strike the enemy as swiftly as a falcon strikes
its target. It surely breaks the back of its prey for the
reason that it awaits the right moment to strike. Its
movement is regulated.*

# 15

Thus the momentum of one skilled in war is
overwhelming, and his attack precisely regulated.[8]

# 16

His potential is that of a fully drawn crossbow;
his timing, the release of the trigger.[9]

# 17

In the tumult and uproar the battle seems chaotic,
but there is no disorder; the troops appear to be
milling about in circles but cannot be defeated.[10]

*Li Ch'üan: In battle all appears to be turmoil and
confusion. But the flags and banners have prescribed arrangements; the sounds of the cymbals,
fixed rules.*

# 18

Apparent confusion is a product of good order;
apparent cowardice, of courage;
apparent weakness, of strength.[11]

*Tu Mu: The verse means that if one wishes to feign*

---

[7] Or regulation of its distance from the prey. [8] Following Tu Mu.

[9] Here again the specific character meaning 'crossbow' is used.

[10] Sun Tzu's onomatopoetic terms suggest the noise and confusion of battle.

[11] Following Tu Mu.

*disorder to entice an enemy 'he must himself be well-disciplined. Only then can he feign confusion. One who wishes to simulate cowardice and lie in wait for his enemy must be courageous, for only then is he able to simulate fear. One who wishes to appear to be weak in order to make his enemy arrogant must be extremely strong. Only then can he feign weakness.*

## 19

Order or disorder depends on organization; courage or cowardice on circumstances; strength or weakness on dispositions.

***Li Ch'üan:*** *Now when troops gain a favourable situation the coward is brave; if it be lost, the brave become cowards. In the art of war there are no fixed rules. These can only be worked out according to circumstances.*

## 20

Thus, those skilled at making the enemy move do so by creating a situation to which he must conform; they entice him with something he is certain to take, and with lures of ostensible profit they await him in strength.

## 21

Therefore a skilled commander seeks victory from the situation and does not demand it of his subordinates.

***Ch'en Hao:*** *Experts in war depend especially on opportunity and expediency. They do not place the burden of accomplishment on their men alone.*

## 22

He selects his men and they exploit the situation.[12]

*Li Ch'üan:* . . . *Now, the valiant can fight; the cautious defend, and the wise counsel. Thus there is none whose talent is wasted.*

*Tu Mu:.* . . *Do not demand accomplishment of those who have no talent.*

When Ts'ao Ts'ao attacked Chang Lu in Han Chung, he left Generals Chang Liao, Li Tien, and Lo Chin in command of over one thousand men to defend Ho Fei. Ts'ao Ts'ao sent instructions to the Army Commissioner, Hsieh Ti, and wrote on the edge of the envelope: 'Open this only when the rebels arrive.' Soon after, Sun Ch'üan of Wu with one hundred thousand men besieged Ho Fei. The generals opened the instructions and read: 'If Sun Ch'üan arrives, Generals Chang and Li will go out to fight. General Lo will defend the city. The Army Commissioner shall not participate in the battle.[13] All the other generals should engage the enemy.'

Chang Liao said: 'Our Lord is campaigning far away, and if we wait for the arrival of rein-forcements the rebels will certainly destroy us. Therefore the instructions say that before the enemy is assembled we should immediately attack him in order to blunt his keen edge and to stabilize the morale of our own troops. Then we can defend the city. The opportunity for victory or defeat lies in this one action.'

Li Tien and Chang Liao went out to attack and actually defeated Sun Ch'üan, and the morale of the Wu army was rubbed out. They returned and put their defences in order and the troops felt secure. Sun

---

[12] The text reads: 'Thus he is able to select men. . .'. That is, men capable of exploiting any situation. A system of selection not based on nepotism or favouritism is the inference.

[13] Ts'ao Ts'ao took care to keep the political officer out of the picture!

141

*Ch'üan assaulted the city for ten days but could not take it and withdrew.*

*The historian Sun Sheng in discussing this observed: 'Now war is a matter of deception. As to the defence of Ho Fei, it was hanging in the air, weak and without reinforcements. If one trusts solely to brave generals who love fighting, this will cause trouble. If one relies solely on those who are cautious, their frightened hearts will find it difficult to control the situation.'*

**Chang Yü:** *Now the method of employing men is to use the avaricious and the stupid, the wise and the brave, and to give responsibility to each in situations that suit him. Do not charge people to do what they cannot do. Select them and give them responsibilities commensurate with their abilities.*

## 23

He who relies on the situation uses his men in fighting as one rolls logs or stones. Now the nature of logs and stones is that on stable ground they are static; on unstable ground, they move. If square, they stop; if round, they roll.

## 24

Thus, the potential of troops skilfully commanded in battle may be compared to that of round boulders which roll down from mountain heights.

**Tu Mu:** *. . . Thus one need use but little strength to achieve much.*

**Chang Yü:** *. . . Li Ching said: 'In war there are three kinds of situation:*

*'When the general is contemptuous of his enemy and his officers love to fight, their ambitions soaring as high as the azure clouds and their spirits as fierce*

*as hurricanes, this is situation in respect to morale.*

'When one man defends a narrow mountain defile which is like sheep's intestines or the door of a dog-house, he can withstand one thousand. This is situation in respect to terrain.

'When one takes advantage of the enemy's laxity, his weariness, his hunger and thirst, or strikes when his advanced camps are not settled, or his army is only half-way across a river, this is situation in respect to the enemy.'

Therefore when using troops, one must take advantage of the situation exactly as if he were setting a ball in motion on a steep slope. The force applied is minute but the results are enormous.

蜀大勝

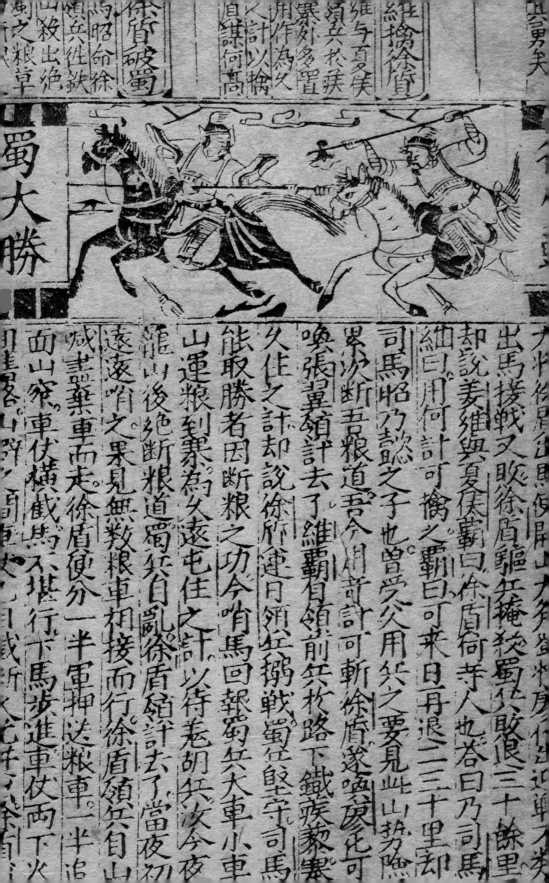

卻說姜維與夏侯霸商議曰徐盾何等人也霸曰乃司馬
昭乃諕之子也曾受父用於之要見此山勢險可
累次斷吾粮道吾今用奇計可斬徐盾遂喚蒙薑
喚張翼領計去了維曰霸領前兵枚路下鐵疾蒙
久住之計卻說徐盾連日領兵稠戰蜀兵堅守司馬
能取勝者因斷粮之功今哨馬回報蜀兵大車小車
山運粮到寨為久遠屯住之計以待差胡兵今夜
隴山後絕斷粮道蜀兵自亂徐盾粮計去了當夜初
遠遠哨之果見無數粮車相接而行徐盾領兵自山
戚盡棄車而走徐盾便分一半軍押送粮車一半迫
面山窄車伏橫截馬不堪行下馬步進車伏兩下火

出馬接戰又敗徐盾驅兵掩殺蜀兵大敗復退三十餘里卻
細曰用何計可擒之霸曰可來日月退二三十里卻

# WEAKNESSES AND STRENGTHS

VI

Sun Tzu said:

## 1

Generally, he who occupies the field of battle first and awaits his enemy is at ease; he who comes later to the scene and rushes into the fight is weary.

## 2

And therefore those skilled in war bring the enemy to the field of battle and are not brought there by him.

## 3

One able to make the enemy come of his own accord does so by offering him some advantage. And one able to prevent him from coming does so by hurting him.

*Tu Yu: . . . If you are able to hold critical points on his strategic roads the enemy cannot come. Therefore Master Wang said: 'When a cat is at the rat hole, ten thousand rats dare not come out; when a tiger guards the ford, ten thousand deer cannot cross.'*

## 4

When the enemy is at ease, be able to weary him; when well fed, to starve him; when at rest, to make him move.

## 5

Appear at places to which he must hasten; move swiftly where he does not expect you.

## 6

That you may march a thousand *li* without wearying yourself is because you travel where there is no enemy.

*Ts'ao Ts'ao: Go into emptiness, strike voids, bypass what he defends, hit him where he does not expect you.*

## 7

To be certain to take what you attack is to attack a place the enemy does not protect. To be certain to hold what you defend is to defend a place the enemy does not attack.

## 8

Therefore, against those skilled in attack, an enemy does not know where to defend; against the experts in defence, the enemy does not know where to attack.

## 9

Subtle and insubstantial, the expert leaves no trace; divinely mysterious, he is inaudible. Thus he is master of his enemy's fate.

*Ho Yen-hsi: . . . I make the enemy see my strengths as weaknesses and my weaknesses as strengths while I cause his strengths to become weaknesses and discover where he is not strong. . . . I conceal my tracks so that none can discern them; I keep silence so that none can hear me.*

## 10

He whose advance is irresistible plunges into his enemy's weak positions; he who in withdrawal cannot be pursued moves so swiftly that he cannot be overtaken.

*Chang Yü: . . . Come like the wind, go like the lightning.*

## 11

When I wish to give battle, my enemy, even though protected by high walls and deep moats, cannot help but engage me, for I attack a position he must succour.

## 12

When I wish to avoid battle I may defend myself simply by drawing a line on the ground; the enemy will be unable to attack me because I divert him from going where he wishes.

*Tu Mu: Chu-ko Liang camped at Yang P'ing and ordered Wei Yen and various generals to combine forces and go down to the east. Chu-ko Liang left only ten thousand men to defend the city while he waited for reports. Ssu-ma I said: 'Chu-ko Liang is in the city; his troops are few; he is not strong. His generals and officers have lost heart.' At this time Chu-ko Liang's spirits were high as usual. He ordered his troops to lay down their banners and silence their drums, and did not allow his men to go out. He opened the four gates and swept and sprinkled the streets.*

*Ssu-ma I suspected an ambush, and led his army in haste to the Northern Mountains.*

*Chu-ko Liang remarked to his Chief of Staff: 'Ssu-ma I thought I had prepared an ambush and*

*fled along the mountain ranges.' Ssu-ma I later*
*learned of this and was overcome with regrets.*[1]

# 13

If I am able to determine the enemy's dispositions
while at the same time I conceal my own then I can
concentrate and he must divide. And if I concentrate
while he divides, I can use my entire strength to attack
a fraction of his.[2] There, I will be numerically superior.

Then, if I am able to use many to strike few at the
selected point, those I deal with will be in dire straits.[3]

***Tu Mu:*** *. . . Sometimes I use light troops and vig-*
*orous horsemen to attack where he is unprepared,*
*sometimes strong crossbowmen and bow-stretching*
*archers to snatch key positions, to stir up his left,*
*overrun his right, alarm him to the front, and strike*
*suddenly into his rear.*

*In broad daylight I deceive him by the use of flags*
*and banners and at night confuse him by beating*
*drums. Then in fear and trembling he will divide*
*his forces to take precautionary measures.*

# 14

The enemy must not know where I intend to give
battle. For if he does not know where I intend to give
battle he must prepare in a great many places. And
when he prepares in a great many places, those I have
to fight in any one place will be few.

---

[1] This story provides the plot for a popular Chinese opera. Chu-ko Liang sat in a gate
tower and played his lute while the porters swept and sprinkled the streets and Ssu-
ma I's host hovered on the outskirts. Ssu-ma I had been fooled before by Chu-ko
Liang and would be fooled again.

[2] Lit. 'one part of his'.

[3] Karlgren GS 1120m for 'dire straits'.

# 15

For if he prepares to the front his rear will be weak, and if to the rear, his front will be fragile. If he prepares to the left, his right will be vulnerable and if to the right, there will be few on his left. And when he prepares everywhere he will be weak everywhere.[4]

*Chang Yü: He will be unable to fathom where my chariots will actually go out, or where my cavalry will actually come from, or where my infantry will actually follow up, and therefore he will disperse and divide and will have to guard against me everywhere. Consequently his force will be scattered and weakened and his strength divided and dissipated, and at the place I engage him I can use a large host against his isolated units.*

# 16

One who has few must prepare against the enemy; one who has many makes the enemy prepare against him.

# 17

If one knows where and when a battle will be fought his troops can march a thousand *li* and meet on the field. But if one knows neither the battleground nor the day of battle, the left will be unable to aid the right, or the right, the left; the van to support the rear, or the rear, the van. How much more is this so when separated by several tens of *li,* or, indeed, by even a few!

*Tu Yü: Now those skilled in war must know where and when a battle will be fought. They measure the roads and fix the date. They divide the army and march in separate columns. Those who are distant*

---

[4] Lit. 'if there is no place he does not make preparations there is no place he is not vulnerable'. The double negative makes the meaning emphatically positive.

*start first, those who are near by, later. Thus the meeting of troops from distances of a thousand* li *takes place at the same time. It is like people coming to a city market.*[5]

## 18

Although I estimate the troops of Yüeh as many, of what benefit is this superiority in respect to the outcome?[6]

## 19

Thus I say that victory can be created. For even if the enemy is numerous, I can prevent him from engaging.
***Chia Lin:*** *Although the enemy be numerous, if he does not know my military situation, I can always make him urgently attend to his own preparations so that he has no leisure to plan to fight me.*

[5] Tu Mu tells the following interesting story to illustrate the point: Emperor Wu of the Sung sent Chu Ling-shih to attack Ch'iao Tsung in Shu. The Emperor Wu said: 'Last year Liu Ching-hsuan went out of the territory inside the river heading for Huang Wu. He achieved nothing and returned. The rebels now think that I should come from outside the river but surmise that I will take them unaware by coming from inside the river. If this is the case they are certain to defend Fu Ch'eng with heavy troops and guard the interior roads. If I go to Huang Wu, I will fall directly into their trap. Now, I will move the main body outside the river and take Ch'eng Tu, and use distracting troops towards the inside of the river. This is a wonderful plan for controlling the enemy.'
Yet he was worried that his plan would be known and that the rebels would learn where he was weak and where strong. So he handed a completely sealed letter to Ling Shih. On the envelope he wrote 'Open when you reach Pai Ti' At this time the army did not know how it was to be divided or from where it would march.
When Ling Shih reached Pai Ti, he opened the letter which read: 'The main body of the army will march together from outside the river to take Ch'eng Tu. Tsang Hsi and Chu Lin from the central river road will take Kuang Han. Send the weak troops embarked in more than ten high boats from within the river toward Huang Wu.' Chiao Tsung actually used heavy troops to defend within the river and Ling Shih exterminated him.
[6] These references to Wu and Yüeh are held by some critics to indicate the date of composition of the text. This point is discussed in the Introduction.

# 20

Therefore, determine the enemy's plans and you
will know which strategy will be successful and which
will not;

# 21

Agitate him and ascertain the pattern of his movement.

# 22

Determine his dispositions and so ascertain the field
of battle.[7]

# 23

Probe him and learn where his strength is abundant
and where deficient.

# 24

The ultimate in disposing one's troops is to be without
ascertainable shape. Then the most penetrating spies
cannot pry in nor can the wise lay plans against you.

# 25

It is according to the shapes that I lay the plans for
victory, but the multitude does not comprehend this.
Although everyone can see the outward aspects, none
understands the way in which I have created victory.

# 26

Therefore, when I have won a victory I do not repeat
my tactics but respond to circumstances in an infinite
variety of ways.

---

[7] Lit. 'the field of life and death'.

## 27

Now an army may be likened to water, for just as
flowing water avoids the heights and hastens to
the lowlands, so an army avoids strength and
strikes weakness.

## 28

And as water shapes its flow in accordance with the
ground, so an army manages its victory in accordance
with the situation of the enemy.

## 29

And as water has no constant form,
there are in war no constant conditions.

## 30

Thus, one able to gain the victory by modifying his
tactics in accordance with the enemy situation
may be said to be divine.

## 31

Of the five elements, none is always predominant;
of the four seasons, none lasts forever; of the days,
some are long and some short, and the moon waxes
and wanes.

# MANOEUVRE[1]

Sun Tzu said:

## 1

Normally, when the army is employed, the general first receives his commands from the sovereign. He assembles the troops and mobilizes the people. He blends the army into a harmonious entity and encamps it.[2]

*Li Ch'üan: He receives the sovereign's mandate and in compliance with the victorious deliberations of the temple councils reverently executes the punishments ordained by Heaven.*

## 2

Nothing is more difficult than the art of manoeuvre. What is difficult about manoeuvre is to make the devious route the most direct and to turn misfortune to advantage.

## 3

Thus, march by an indirect route and divert the enemy by enticing him with a bait. So doing, you may set out after he does and arrive before him. One able to

---

[1] Lit. 'struggle' or 'contest of the armies' as each strives to gain an advantageous position. [2] This verse can be translated as I have, following Li Ch'üan and Chia Lin, or 'He encamps the army so that the Gates of Harmony confront one another' following Ts'ao Ts'ao and Tu Mu. After assembling the army the first task of a commander would be to organize it, or to 'harmonize' its diverse elements.

do this understands the strategy of the direct and the indirect.

*Ts'ao Ts'ao: . . . Make it appear that you are far off. You may start after the enemy and arrive before him because you know how to estimate and calculate distances.*

*Tu Mu:*[3] *He who wishes to snatch an advantage takes a devious and distant route and makes of it the short way. He turns misfortune to his advantage. He deceives and fools the enemy to make him dilatory and lax, and then marches on speedily.*

## 4

Now both advantage and danger are inherent in manoeuvre.[4]

*Ts'ao Ts'ao: One skilled will profit by it; if he is not, it is dangerous.*

## 5

One who sets the entire army in motion to chase an advantage will not attain it.

## 6

If he abandons the camp to contend for advantage the stores will be lost.

*Tu Mu: If one moves with everything the stores will travel slowly and he will not gain the advantage.*

---

[3] This comment appears under v. 2 in the text.

[4] Giles based his reading on the TT and translated: 'Manoeuvring with an army is advantageous; with an undisciplined multitude most dangerous.' Sun Hsing-yen also thought this was the meaning of the verse. This too literal translation completely misses the point. Ts'ao Ts'ao's interpretation is surely more satisfactory. The verse is a generalization which introduces what follows. A course of action which may appear advantageous usually contains within itself the seeds of disadvantage. The converse is also true.

*If he leaves the heavy baggage behind and presses on with the light troops, it is to be feared the baggage would be lost.*

7

It follows that when one rolls up the armour and sets out speedily, stopping neither day nor night and marching at double time for a hundred *li,* the three commanders will be captured. For the vigorous troops will arrive first and the feeble straggle along behind, so that if this method is used only one-tenth of the army will arrive.[5]

*Tu Mu: . . . Normally, an army marches thirty* li *in a day, which is one stage. In a forced march of double distance it covers two stages. You can cover one hundred* li *only if you rest neither day nor night. If the march is conducted in this manner the troops will be taken prisoner. . . . When Sun Tzu says that if this method is used only one out of ten will arrive he means that when there is no alternative and you must contend for an advantageous position, you select the most robust man of ten to go first while you order the remainder to follow in the rear. So of ten thousand men you select one thousand who will arrive at dawn. The remainder will arrive continuously, some in late morning and some in mid-afternoon, so that none is exhausted and all arrive in succession to join those who preceded them. The sound of their marching is uninterrupted. In contending for advantage, it must be for a strategically critical point. Then, even one thousand will be sufficient to defend it until those who follow arrive.*

---

[5] By 'rolling up armour' Sun Tzu undoubtedly meant that heavy individual equipment would be bundled together and left at base.

# 8

In a forced march of fifty *li* the commander of the van will fall, and using this method but half the army will arrive. In a forced march of thirty *li,* but two-thirds will arrive.[6]

# 9

It follows that an army which lacks heavy equipment, fodder, food and stores will be lost.[7]

**Li Ch'üan:** . . . *The protection of metal walls is not as important as grain and food.*

# 10

Those who do not know the conditions of mountains and forests, hazardous defiles, marshes and swamps, cannot conduct the march of an army;

# 11

Those who do not use local guides are unable to obtain the advantages of the ground.

**Tu Mu:** *The* Kuan Tzu *says: 'Generally, the commander must thoroughly acquaint himself beforehand with the maps so that he knows dangerous places for chariots and carts, where the water is too deep for wagons; passes in famous mountains,[8] the principal rivers, the locations of highlands and hills;*

---

[6] This may also be rendered as 'The general of the Upper Army [as distinguished from the generals commanding the Central and Lower Armies] will be defeated' or 'will be checked'. Here the Upper Army would refer to the advance guard when the three divisions of the army marched in column. In other words, the advantages and disadvantages of forced marches must be carefully weighed, and the problem of what should be carried and what left in a secure base considered.

[7] The verse which follows this one repeats a previous verse and is a *non sequitur* here. It has been dropped.

[8] 'Famous' because of their strategic significance.

*where rushes, forests, and reeds are luxuriant; the road distances; the size of cities and towns; well-known cities and abandoned ones, and where there are flourishing orchards. All this must be known, as well as the way boundaries run in and out. All these facts the general must store in his mind; only then will he not lose the advantage of the ground.'*

*Li Ching said: '. . . We should select the bravest officers and those who are most intelligent and keen, and using local guides, secretly traverse mountain and forest noiselessly and concealing our traces. Sometimes we make artificial animals' feet to put on our feet; at others we put artificial birds on our hats and quietly conceal ourselves in luxuriant undergrowth. After this, we listen carefully for distant sounds and screw up our eyes to see clearly. We concentrate our wits so that we may snatch an*

*opportunity. We observe the indications of the atmosphere; look for traces in the water to know if the enemy has waded a stream, and watch for movement of the trees which indicates his approach.'*

**Ho Yen-hsi:** *. . . Now, if having received instructions to launch a campaign, we hasten to unfamiliar land where cultural influence has not penetrated and communications are cut, and rush into its defiles, is it not difficult? If I go with a solitary army the enemy awaits me vigilantly. For the situations of an attacker and a defender are vastly different. How much more so when the enemy concentrates on deception and uses many misleading devices! If we have made no plans we plunge in headlong. By braving the dangers and entering perilous places we face the calamity of being trapped or inundated. Marching as if drunk, we may run into an unexpected fight. When we stop at night we are worried by false alarms; if we hasten along unprepared we fall into ambushes. This is to plunge an army of bears and tigers into the land of death. How can we cope with the rebels' fortifications, or sweep him out of his deceptive dens?*

*Therefore in the enemy's country, the mountains, rivers, highlands, lowlands and hills which he can defend as strategic points; the forests, reeds, rushes and luxuriant grasses in which he can conceal himself; the distances over the roads and paths, the size of cities and towns, the extent of the villages, the fertility or barrenness of the fields, the depth of irrigation works, the amounts of stores, the size of the opposing army, the keenness of weapons—all must be fully known. Then we have the enemy in our sights and he can be easily taken.*

# 12

Now war is based on deception. Move when it is advantageous and create changes in the situation by dispersal and concentration of forces.[9]

# 13

When campaigning, be swift as the wind; in leisurely march, majestic as the forest; in raiding and plundering, like fire; in standing, firm as the mountains.[10] As unfathomable as the clouds, move like a thunderbolt.

# 14

When you plunder the countryside, divide your forces.[11] When you conquer territory, divide the profits.[12]

# 15

Weigh the situation, then move.

# 16

He who knows the art of the direct and the indirect approach will be victorious. Such is the art of manoeuvring.

# 17

The Book of Military Administration says: 'As the voice cannot be heard in battle, drums and bells are used. As troops cannot see each other clearly in battle, flags and banners are used.'[13]

---

[9] Mao Tse-tung paraphrases this verse several times.

[10] Adopted as his slogan by the Japanese warrior Takeda Shingen.

[11] Yang P'ing-an emends and reads: 'Thus wherever your banners point, the enemy is divided.' There does not seem to be any justification for this change.

[12] Rather than 'divide the profits' Yang P'ing-an reads: 'defend it to the best advantage'. The text does not substantiate this rendering.

[13] This verse is interesting because in it Sun Tzu names a work which antedates his own.

# 18

Now gongs and drums, banners and flags are used to focus the attention of the troops. When the troops can be thus united, the brave cannot advance alone, nor can the cowardly withdraw. This is the art of employing a host.

**Tu Mu:**. . . *The Military Law states: 'Those who when they should advance do not do so and those who when they should retire do not do so are beheaded.'*

*When Wu Ch'i fought against Ch'in, there was an officer who before battle was joined was unable to control his ardour. He advanced and took a pair of heads and returned, Wu Ch'i ordered him beheaded. The Army Commissioner admonished him, saying: 'This is a talented officer; you should not behead him.' Wu Ch'i replied: 'I am confident he is an officer of talent, but he is disobedient.' Thereupon he beheaded him.*

# 19

In night fighting use many torches and drums, in day fighting many banners and flags in order to influence the sight and hearing of our troops.[14]

**Tu Mu:** . . *Just as large formations include smaller ones, so large camps include smaller ones. The army of the van, rear, right and left has each its own camp. These form a circle round the headquarters of the commander-in-chief in the centre. All the camps encompass the headquarters. The several corners are hooked together so that the camp appears like the* Pi Lei *constellation.*[15]

*The distance between camps is not greater than one hundred paces or less than fifty. The roads and paths join to enable troops to parade. The fortifications face each other so that each can assist the others with bows and crossbows.*

*At every crossroad a small fort is built; on top firewood is piled; inside there are concealed tunnels. One climbs up to these by ladders; sentries are*

---

[14] Or 'the enemy', it is not clear which. Possibly both. Tu Mu's comment is not particularly relevant to the verse but is included because it indicates a remarkably high degree of skill in the science of castramentation.

[15] Markal? *Pi* is Alpharatz.

*stationed there. After darkness, if a sentry hears drumbeats on the four sides of the camp he sets off the beacon fire. Therefore if the enemy attacks at night he may get in at the gates, but everywhere there are small camps, each firmly defended, and to the east, west, north or south he does not know which to attack.*

*In the camp of the commander-in-chief or in the smaller camps, those who first know the enemy has come allow them all to enter; they then beat the drums and all the camps respond. At all the small forts beacon fires are lit, making it as light as day. Whereupon the officers and troops close the gates of the camps and man the fortifications and look down*

*upon the enemy. Strong crossbows and powerful bows shoot in all directions. . . .*

*Our only worry is that the enemy will not attack at night, for if he does he is certain to be defeated.*

# 20

Now an army may be robbed of its spirit and its commander deprived of his courage.[16]

***Ho Yen-hsi:*** *. . . Wu Ch'i said: 'The responsibility for a martial host of a million lies in one man. He is the trigger of its spirit.'*

---

[16] Or 'of his wits', I am not sure which.

*Mei Yao-ch'en:* . . . *If an army has been deprived of its morale, its general will also lose his heart.*

*Chang Yü: Heart is that by which the general masters. Now order and confusion, bravery and cowardice, are qualities dominated by the heart. Therefore the expert at controlling his enemy frustrates him and then moves against him. He aggravates him to confuse him and harasses him to make him fearful. Thus he robs his enemy of his heart and of his ability to plan.*

# 21

During the early morning spirits are keen, during the day they flag, and in the evening thoughts turn toward home.[17]

# 22

And therefore those skilled in war avoid the enemy when his spirit is keen and attack him when it is sluggish and his soldiers homesick. This is control of the moral factor.

# 23

In good order they await a disorderly enemy; in serenity, a clamorous one. This is control of the mental factor.

*Tu Mu: In serenity and firmness they are not destroyed by events.*

*Ho Yen-hsi: For the lone general who with subtlety must control a host of a million against an enemy as fierce as tigers, advantages and disadvantages are intermixed. In the face of countless changes he must*

---

[17] Mei Yao-ch'en says that 'morning', 'day', and 'evening' represent the phases of a long campaign.

*be wise and flexible; he must bear in mind all
possibilities. Unless he is stout of heart and his
judgement not confused, how would he be able to
respond to circumstances without coming to his wits'
end? And how settle affairs without being bewildered? When unexpectedly confronted with grave
difficulties, how could he not be alarmed?
How could he control the myriad matters without
being confused?*

混
乱
的

# 24

Close to the field of battle, they await an enemy
coming from afar; at rest, an exhausted enemy; with
well-fed troops, hungry ones. This is control of the
physical factor.

# 25

They do not engage an enemy advancing with
well-ordered banners nor one whose formations are
in impressive array. This is control of the factor of
changing circumstances.[18]

# 26

Therefore, the art of employing troops is that when the
enemy occupies high ground, do not confront him;
with his back resting on hills, do no oppose him.

# 27

When he pretends to flee, do not pursue.

# 28

Do not attack his *élite* troops.

---

[18] Or the 'circumstantial factor'. 'They' in these verses refers to those skilled in war.

# 29

Do not gobble proferred baits.

*Mei Yao-ch'en: The fish which covets bait is caught; troops who covet bait are defeated.*

*Chang Yü: The 'Three Strategies' says: 'Under fragrant bait there is certain to be a hooked fish.'*

# 30

Do not thwart an enemy returning homewards.

# 31

To a surrounded enemy you must leave a way of escape.

*Tu Mu: Show him there is a road to safety, and so create in his mind the idea that there is an alternative to death. Then strike.*

*Ho Yen-hsi: When Ts'ao Ts'ao surrounded Hu Kuan he issued an order: 'When the city is taken, the defenders will be buried.' For month after month it did not fall. Ts'ao Jen said: 'When a city is surrounded it is essential to show the besieged that there is a way to survival. Now, Sir, as you have told them they must fight to the death everyone will fight to save his own skin. The city is strong and has a plentiful supply of food. If we attack them many officers and men will be wounded. If we persevere in this it will take many days. To encamp under the walls of a strong city and attack rebels determined to fight to the death is not a good plan!' Ts'ao Ts'ao followed this advice, and the city submitted.*

## 32

Do not press an enemy at bay.

*Tu Yu: Prince Fu Ch'ai said: 'Wild beasts, when at bay, fight desperately. How much more is this true of men! If they know there is no alternative they will fight to the death.'*

*During the reign of Emperor Hsüan of the Han, Chao Ch'ung-kuo was suppressing a revolt of the Ch'iang tribe. The Ch'iang tribesmen saw his large army, discarded their heavy baggage, and set out to ford the Yellow River. The road was through narrow defiles, and Ch'ung Kuo drove them along in a leisurely manner.*

*Someone said: 'We are in pursuit of great advantage but proceed slowly.'*

*Ch'ung-kuo replied: 'They are desperate. I cannot press them. If I do this easily they will go without even looking around. If I press them they will turn on us and fight to the death.'*

*All the generals said: 'Wonderful!'*

## 33

This is the method of employing troops.

# THE NINE VARIABLES

Sun Tzu said:

## 1

In general, the system of employing troops is that the commander receives his mandate from the sovereign to mobilize the people and assemble the army.[1]

## 2

You should not encamp in low-lying ground.

## 3

In communicating ground, unite with your allies.

## 4

You should not linger in desolate ground.

## 5

In enclosed ground, resourcefulness is required.

## 6

In death ground, fight.

---

[1] As Sun Tzu uses almost identical words to introduce chapter vii, Yang Ping-an would drop this. He would also drop v. 2–6 inclusive, as they occur later in discussion of the 'Nine Grounds', and replace them with v. 26–32 inclusive from chapter vii. Where Sun Tzu uses a negative in v. 2–6 it is not the peremptory form he used previously. Hence I do not feel justified in accepting the emendations proposed. The 'Nine Variables' are then expressed in v. 2–7 inclusive.

## 7

There are some roads not to follow; some troops not to strike; some cities not to assault; and some ground which should not be contested.

*Wang Hsi: In my opinion, troops put out as bait, élite troops, and an enemy in well-regulated and imposing formation should not be attacked.*

*Tu Mu: Probably this refers to an enemy in a strategic position behind lofty walls and deep moats with a plentiful store of grain and food, whose purpose is to detain my army. Should I attack the city and take it, there would be no advantage worth mentioning; if I do not take it the assault will certainly grind down the power of my army. Therefore I should not attack it.*

## 8

There are occasions when the commands of the sovereign need not be obeyed.[2]

*Ts'ao Ts'ao: When it is expedient in operations the general need not be restricted by the commands of the sovereign.*

*Tu Mu: The* Wei Liao Tzu *says: 'Weapons are inauspicious instruments; strife contrary to virtue; the general, the Minister of Death, who is not responsible to the heavens above, to the earth beneath, to the enemy in his front, or to the sovereign in his rear.'*

*Chang Yü: Now King Fu Ch'ai said: 'When you see the correct course, act; do not wait for orders.'*

## 9

A general thoroughly versed in the advantages of the nine variable factors knows how to employ troops.

*Chia Lin: The general must rely on his ability to control the situation to his advantage as opportunity dictates. He is not bound by established procedures.*

## 10

The general who does not understand the advantages of the nine variable factors will not be able to use the ground to his advantage even though familiar with it.

*Chia Lin: . . . A general prizes opportune changes in circumstances.*

## 11

In the direction of military operations one who does not understand the tactics suitable to the nine variable situations will be unable to use his troops effectively, even if he understands the 'five advantages'.[3]

*Chia Lin:. . . The 'five variations' are the following: A road, although it may be the shortest, is not to be followed if one knows it is dangerous and there is the contingency of ambush.*

*An army, although it may be attacked, is not to be attacked if it is in desperate circumstances and there is the possibility that the enemy will fight to the death.*

*A city, although isolated and susceptible to attack, is not to be attacked if there is the probability that it is well stocked with provisions, defended by crack troops under command of a wise general, that its ministers are loyal and their plans unfathomable.*

*Ground, although it may be contested, is not to be fought for if one knows that after getting it, it*

---

[2] A catch-all which covers the variable circumstances previously enumerated.

[3] A confusing verse which baffles all the commentators. If Chia Lin is correct the 'five advantages' must be the situations named in v. 2–6 inclusive.

*will be difficult to defend, or that he gains no advantage by obtaining it, but will probably be counter-attacked and suffer casualties.*

*The orders of a sovereign, although they should be followed, are not to be followed if the general knows they contain the danger of harmful superintendence of affairs from the capital.*

*These five contingencies must be managed as they arise and as circumstances dictate at the time, for they cannot be settled beforehand.*

# 12

And for this reason, the wise general in his deliberations must consider both favourable and unfavourable factors.[4]

*Ts'ao Ts'ao: He ponders the dangers inherent in the advantages, and the advantages inherent in the dangers.*

# 13

By taking into account the favourable factors, he makes his plan feasible; by taking into account the unfavourable, he may resolve the difficulties.[5]

*Tu Mu:. . . If I wish to take advantage of the enemy I must perceive not just the advantage in doing so but must first consider the ways he can harm me if I do.*

*Ho Yen-hsi: Advantage and disadvantage are mutually reproductive. The enlightened deliberate.*

---

[4] Sun Tzu says these are 'mixed'.

[5] Sun Tzu says that by taking account of the favourable factors the plan is made 'trustworthy' or 'reliable'. 'Feasible' (or 'workable') is as close as I can get it.

# 14

He who intimidates his neighbours does so by
inflicting injury upon them.

*Chia Lin: Plans and projects for harming the
enemy are not confined to any one method.
Sometimes entice his wise and virtuous men away
so that he has no counsellors. Or send treacherous
people to his country to wreck his administration.
Sometimes use cunning deceptions to alienate his
ministers from the sovereign. Or send skilled craftsmen to encourage his people to exhaust their wealth.
Or present him with licentious musicians and
dancers to change his customs. Or give him beautiful women to bewilder him.*

# 15

He wearies them by keeping them constantly occupied,
and makes them rush about by offering them
ostensible advantages.

# 16

It is a doctrine of war not to assume the enemy will
not come, but rather to rely on one's readiness to meet
him; not to presume that he will not attack, but rather
to make one's self invincible.

*Ho Yen-hsi: . . . The 'Strategies of Wu' says: 'When
the world is at peace, a gentleman keeps his sword
by his side.'*

# 17

There are five qualities which are dangerous in the
character of a general.

## 18

If reckless, he can be killed;

**Tu Mu:** *A general who is stupid and courageous is a calamity. Wu Ch'i said: 'When people discuss a general they always pay attention to his courage. As far as a general is concerned, courage is but one quality. Now a valiant general will be certain to enter an engagement recklessly and if he does so he will not appreciate what is advantageous.'*

## 19

If cowardly, captured;

**Ho Yen-hsi:** *The* Ssu-ma Fa *says: 'One who esteems life above all will be overcome with hesitancy. Hesitancy in a general is a great calamity.'*

## 20

If quick-tempered you can make a fool of him;

**Tu Yu:** *An impulsive man can be provoked to rage and brought to his death. One easily angered is irascible, obstinate and hasty. He does not consider difficulties.*

**Wang Hsi:** *What is essential in the temperament of a general is steadiness.*

## 21

If he has too delicate a sense of honour you can
calumniate him;

*Mei Yao-ch'en: One anxious to defend his reputa-
tion pays no regard to anything else.*

## 22

If he is of a compassionate nature you can harass him.

*Tu Mu: He who is humanitarian and compassion-
ate and fears only casualties cannot give up temporary advantage for a long-term gain and is unable
to let go this in order to seize that.*

## 23

Now these five traits of character are serious faults in
a general and in military operations are calamitous.

## 24

The ruin of the army and the death of the general are
inevitable results of these shortcomings. They must be
deeply pondered.

# MARCHES

Sun Tzu said:

## 1

Generally when taking up a position and confronting the enemy, having crossed the mountains, stay close to valleys. Encamp on high ground facing the sunny side.[1]

## 2

Fight downhill; do not ascend to attack.[2]

## 3

So much for taking position in mountains.

## 4

After crossing a river you must move some distance away from it.

## 5

When an advancing enemy crosses water do not meet him at the water's edge. It is advantageous to allow half his force to cross and then strike.

*Ho Yen-hsi: During the Spring and Autumn period the Duke of Sung came to Hung to engage the Ch'u*

---

[1] Lit. 'Looking in the direction of growth, camp in a high place.' The commentators explain that *sheng* (生), 'growth', means *yang* (陽), 'sunny'—i.e. the south.

[2] The TT reading is adopted. Otherwise: 'In mountain warfare, do not ascend to attack.'

*army. The Sung army had deployed before the Ch'u troops had completed crossing the river. The Minister of War said: 'The enemy is many, we are but few. I request permission to attack before he has completed his crossing.' The Duke of Sung replied: 'You cannot.'*

*When the Ch'u army had completed the river crossing but had not yet drawn up its formations, the Minister again asked permission to attack and the Duke replied: 'Not yet. When they have drawn up their army we will attack.'*

*The Sung army was defeated, the Duke wounded in the thigh, and the officers of the Van annihilated.[3]*

# 6

If you wish to give battle, do not confront your enemy close to the water.[4] Take position on high ground facing the sunlight. Do not take position downstream.

# 7

This relates to taking up positions near a river.

# 8

Cross salt marshes speedily. Do not linger in them. If you encounter the enemy in the middle of a salt marsh you must take position close to grass and water with trees to your rear.[5]

# 9

This has to do with taking up position in salt marshes.

---

[3] The source of Mao Tse-tung's remark: 'We are not like the Duke of Sung.'

[4] The commentators say that the purpose of retiring from the banks or shores is to lure the enemy to attempt a crossing.

[5] Possibly salt flats from time to time inundated, as one sees in north and east China, rather than the salt marshes negotiable only by boat, with which we are more familiar.

# 10

In level ground occupy a position which facilitates your action. With heights to your rear and right, the field of battle is to the front and the rear is safe.[6]

# 11

This is how to take up position in level ground.

# 12

Generally, these are advantageous for encamping in the four situations named.[7] By using them the Yellow Emperor conquered four sovereigns.[8]

# 13

An army prefers high ground to low; esteems sunlight and dislikes shade. Thus, while nourishing its health, the army occupies a firm position. An army that does not suffer from countless diseases is said to be certain of victory.[9]

# 14

When near mounds, foothills, dikes or embankments, you must take position on the sunny side and rest your right and rear on them.

# 15

These methods are all advantageous for the army, and gain the help the ground affords.[10]

---

[6] Sun Tzu says: 'To the front, death; to the rear, life.' The right flank was the more vulnerable; shields were carried on the left arm.

[7] That is, the methods described are to be used in encamping the army. Chang Yü takes the verses to relate to encamping but then proceeds to quote Chu-ko Liang on fighting in such places.

[8] Supposed to have reigned 2697–2597 BC.

[9] Lit. 'the one hundred diseases'. [10] Following page.

# 16

Where there are precipitous torrents, 'Heavenly Wells',
'Heavenly Prisons', 'Heavenly Nets', 'Heavenly Traps',
and 'Heavenly Cracks', you must march speedily away
from them. Do not approach them.

*Ts'ao Ts'ao: Raging waters in deep mountains are
'precipitous torrents'. A place surrounded by heights
with low-lying ground in the centre is called a
'Heavenly Well'. When you pass through mountains
and the terrain resembles a covered cage it is a
'Heavenly Prison'. Places where troops can be
entrapped and cut off are called 'Heavenly Nets'.
Where the land is sunken, it is a 'Heavenly Trap'.
Where mountain gorges are narrow and where the
road is sunken for several tens of feet, this is a
'Heavenly Crack'.*

# 17

I keep a distance from these and draw the enemy
toward them. I face them and cause him
to put his back to them.

# 18

When on the flanks of the army there are dangerous
defiles or ponds covered with aquatic grasses where
reeds and rushes grow, or forested mountains with
dense tangled undergrowth you must carefully search
them out, for these are places where ambushes are
laid and spies are hidden.

---

[10] The verse which immediately follows this in the text reads: 'When rain falls in the
upper reaches of a river and foaming water descends those who wish to ford must
wait until the waters subside.' This is obviously out of place here. I suspect it is part
of the commentary which has worked its way into the text.

## 19

When the enemy is near by but lying low he is
depending on a favourable position. When he
challenges to battle from afar he wishes to lure you
to advance, for when he is in easy ground
he is in an advantageous position.[11]

## 20

When the trees are seen to move the enemy
is advancing.

[11] Another version seems to have read: '. . . is offering an ostensible advantage.'

## 21

When many obstacles have been placed in the undergrowth, it is for the purpose of deception.

## 22

Birds rising in flight is a sign that the enemy is lying in ambush; when the wild animals are startled and flee he is trying to take you unaware.

## 23

Dust spurting upward in high straight columns indicates the approach of chariots. When it hangs low and is widespread infantry is approaching.

*Tu Mu: When chariots and cavalry travel rapidly they come one after another like fish on a string and therefore the dust rises high in slender columns.*

*Chang Yü: . . . Now when the army marches there should be patrols out to the front to observe. If they see dust raised by the enemy, they must speedily report this to the commanding general.*

## 24

When dust rises in scattered areas the enemy is bringing in firewood; when there are numerous small patches which seem to come and go he is encamping the army.[12]

## 25

When the enemy's envoys speak in humble terms, but he continues his preparations, he will advance.

*Chang Yü: When T'ien Tan was defending Chi Mo*

---

[12] Li Ch'üan's reading, 'bringing in firewood', is adopted. They are dragging bundles of firewood. The comments that interrupt this verse are devoted to discussions of how people collect firewood!

*the Yen general Ch'i Che surrounded it. T'ien Tan personally handled the spade and shared in the labour of the troops. He sent his wives and concubines to enroll in the ranks and divided his own food to entertain his officers. He also sent women to the city walls to ask for terms of surrender. The Yen general was very pleased. T'ien Tan also collected twenty-four thousand ounces of gold, and made the rich citizens send a letter to the Yen general which said: 'The city is to be surrendered immediately. Our only wish is that you will not make our wives and concubines prisoners.' The Yen army became increasingly relaxed and negligent and T'ien Tan sallied out of the city and inflicted a crushing defeat on them.*

## 26

When their language is deceptive but the enemy pretentiously advances, he will retreat.

## 27

When the envoys speak in apologetic terms, he wishes a respite.[13]

## 28

When without a previous understanding the enemy asks for a truce, he is plotting.

*Ch'ên Hao:. . . If without reason one begs for a truce it is assuredly because affairs in his country are in a dangerous state and he is worried and wishes to make a plan to gain a respite. Or otherwise he knows that our situation is susceptible to his*

---

[13] This verse, out of place in the text, has been moved to the present context.

*plots and he wants to forestall our suspicions by
asking for a truce. Then he will take advantage
of our unpreparedness.*

## 29

When light chariots first go out and take position
on the flanks the enemy is forming for battle.
*Chang Yü: In the 'Fish Scale Formation' chariots
are in front, infantry behind them.*

## 30

When his troops march speedily and he parades
his battle chariots he is expecting to rendezvous
with reinforcements.[14]

# 31

When half his force advances and half withdraws
he is attempting to decoy you.

# 32

When his troops lean on their weapons,
they are famished.

# 33

When drawers of water drink before carrying it
to camp, his troops are suffering from thirst.

---

[14] This is not exactly clear. He expects to rendezvous with reinforcing troops?
Or are his dispersed detachments concentrating?

# 34

When the enemy sees an advantage but does not advance to seize it, he is fatigued.[15]

# 35

When birds gather above his camp sites,
they are empty.

*Ch'en Hao: Sun Tzu is describing how to distinguish between the true and the false in the enemy's aspect.*

# 36

When at night the enemy's camp is clamorous,
he is fearful.[16]

*Tu Mu: His troops are terrified and insecure. They are boisterous to reassure themselves.*

# 37

When his troops are disorderly, the general has no prestige.

*Ch'en Hao: When the general's orders are not strict and his deportment undignified, the officers will be disorderly.*

# 38

When his flags and banners move about constantly he is in disarray.

*Tu Mu: Duke Chuang of Lu defeated Ch'i at Ch'ang Sho. Tsao Kuei requested permission to pursue. The Duke asked him why. He replied: 'I see that the ruts of their chariots are confused and their*

---

[15] The fact that this series of verses is expressed in elementary terms does not restrain the commentators, who insist on explaining each one at considerable length.
[16] See Plutarch's description in 'Alexander' of the Persian camp the night before the battle of Gaugemala.

*flags and banners drooping. Therefore I wish to pursue them.'*

## 39

If the officers are short-tempered they are exhausted.

**Ch'ên Hao:** *When the general lays on unnecessary projects, everyone is fatigued.*

**Chang Yü:** *When administration and orders are inconsistent, the men's spirits are low, and the officers exceedingly angry.*

## 40

When the enemy feeds grain to the horses and his men meat and when his troops neither hang up their cooking pots nor return to their shelters, the enemy is desperate.[17]

**Wang Hsi:** *The enemy feeds grain to the horses and the men eat meat in order to increase their strength and powers of endurance. If the army has no cooking pots it will not again eat. If the troops do not go back to their shelters they have no thoughts of home and intend to engage in decisive battle.*

## 41

When the troops continually gather in small groups and whisper together the general has lost the confidence of the army.[18]

---

[17] Chang Yü says that when an army 'burns its boats' and 'smashes its cooking pots' it is at bay and will fight to the death.

[18] Comments under this verse are principally devoted to explaining the terms used. Most of the commentators agree that when the men gather together and carry on whispered conversations they are criticizing their officers. Mei Yaoch'en observes that they are probably planning to desert. The verse which immediately follows is a paraphrase of this one, and is dropped.

# 42

Too frequent rewards indicate that the general is
at the end of his resources; too frequent punishments
that he is in acute distress.[19]

# 43

If the officers at first treat the men violently and later
are fearful of them, the limit of indiscipline has been
reached.[20]

# 44

When the enemy troops are in high spirits, and,
although facing you, do not join battle for a long time,
nor leave, you must thoroughly investigate the situation.

# 45

In war, numbers alone confer no advantage.
Do not advance relying on sheer military power.[21]

# 46

It is sufficient to estimate the enemy situation correctly
and to concentrate your strength to capture him.[22]
There is no more to it than this. He who lacks
foresight and underestimates his enemy will surely
be captured by him.

# 47

If troops are punished before their loyalty is secured
they will be disobedient. If not obedient, it is difficult
to employ them. If troops are loyal, but punishments
are not enforced, you cannot employ them.

# 48

Thus, command them with civility and imbue them
uniformly with martial ardour and it may be said that
victory is certain.

# 49

If orders which are consistently effective are used in
instructing the troops, they will be obedient. If orders
which are not consistently effective are used in
instructing them, they will be disobedient.

# 50

When orders are consistently trustworthy and
observed, the relationship of a commander
with his troops is satisfactory.

---

[19] Ho Yen-hsi remarks that in the management of his affairs the general should
seek a balance of tolerance and severity.

[20] Or 'at first to bluster but later to be in fear of the enemy's host'? Ts'ao Ts'ao,
Tu Mu, Wang Hsi and Chang Yü all take the *ch'i* (其) here to refer to the enemy,
but this thought does not follow the preceding verse too well. Tu Yu's interpretation,
which I adopt, seems better.

[21] 'For it is not by the numbers of the combatants but by their orderly array
and their bravery that prowess in war is wont to be measured.'
Procopius, *History of the Wars*, p.347.

[22] Ts'ao Ts'ao misinterprets *tsu* (足) here in the phrase *tsu i* (足以) meaning
'it is sufficient'. His mistake obviously confused the commentators and none cared
to take issue with him. Wang Hsi starts off bravely enough by saying: 'I think those
who are skilled in creating changes in the situation by concentration and dispersion
need only gather their forces together and take advantage of a chink in the enemy's
defenses to gain the victory', but in the end allows Ts'ao Ts'ao's prestige
to overcome his own better judgement.

# TERRAIN[1]

Sun Tzu said:

## 1

Ground may be classified according to its nature as accessible, entrapping, indecisive, constricted, precipitous, and distant.[2]

## 2

Ground which both we and the enemy can traverse with equal ease is called accessible. In such ground, he who first takes high sunny positions convenient to his supply routes can fight advantageously.

## 3

Ground easy to get out of but difficult to return to is entrapping. The nature of this ground is such that if the enemy is unprepared and you sally out you may defeat him. If the enemy is prepared and you go out and engage, but do not win, it is difficult to return. This is unprofitable.

---

[1] 'Topography' or 'conformation of the ground'.

[2] Mei Yao-ch'en defines 'accessible' ground as that in which roads meet and cross; 'entrapping' ground as net-like; 'indecisive' ground as that in which one gets locked with the enemy; 'constricted' ground as that in which a valley runs between two mountains; 'precipitous' ground as that in which there are mountains, rivers, foothills and ridges, and 'distant' ground as level. Sun Tzu uses 'distant' to indicate that there is a considerable distance between the camps of the two armies.

# 4

Ground equally disadvantageous for both the enemy and ourselves to enter is indecisive. The nature of this ground is such that although the enemy holds out a bait I do not go forth but entice him by marching off. When I have drawn out half his force, I can strike him advantageously.

*Chang Yü:. . . Li Ching's 'Art of War' says: 'In ground which offers no advantage to either side we should lure the enemy by feigned departure, wait until half his force has come out, and make an intercepting attack.'*

# 5

If I first occupy constricted ground I must block the passes and await the enemy. If the enemy first occupies such ground and blocks the defiles I should not follow him; if he does not block them completely I may do so.

# 6

In precipitous ground I must take position on the sunny heights and await the enemy.[3] If he first occupies such ground I lure him by marching off; I do not follow him.

*Chang Yü: If one should be the first to occupy a position in level ground, how much more does this apply to difficult and dangerous places![4] How can such terrain be given to the enemy?*

---

[3] Generally I have translated the *Yang* of *Yin-Yang* as 'south' or 'sunny', and *Yin* as 'north' or 'shady'. In the context of Sun Tzu these terms have no cosmic connotations.

[4] *Hsien* (險) means a 'narrow pass', hence 'dangerous' and by implication 'strategic'.

## 7

When at a distance from an enemy of equal strength it is difficult to provoke battle and unprofitable to engage him in his chosen position.[5]

## 8

These are the principles relating to six different types of ground. It is the highest responsibility of the general to inquire into them with the utmost care.

*Mei Yao-ch'en: Now the nature of the ground is the fundamental factor in aiding the army to set up its victory.*

## 9

Now when troops flee, are insubordinate,[6] distressed, collapse in disorder or are routed, it is the fault of the general. None of these disasters can be attributed to natural causes.

## 10

Other conditions being equal, if a force attacks one ten times its size, the result is flight.

*Tu Mu: When one is to be used to attack ten we should first compare the wisdom and the strategy of the opposing generals, the bravery and cowardice of the troops, the question of weather, of the advantages offered by the ground, whether the troops are well fed or hungry, weary or rested.*

---

5 The phrase following 'engage' is added to clarify Sun Tzu's meaning.

6 The character rendered 'insubordinate' is *shih* (弛), 'to unstring a bow'; hence, 'lax', 'remiss', 'loose'. The commentators make it clear that in this context the character means 'insubordinate'.

# 11

When troops are strong and officers weak
the army is insubordinate.

*Tu Mu: The verse is speaking of troops and non-
commissioned officers[7] who are unruly and overbearing, and generals and commanders who are
timid and weak.*

*In the present dynasty at the beginning of the
Ch'ang Ch'ing reign period[8] T'ien Pu was ordered
to take command in Wei for the purpose of attacking Wang T'ing-ch'ou. Pu had grown up in Wei and
the people there held him in contempt, and several
tens of thousands of men all rode donkeys about the
camp. Pu was unable to restrain them. He
remained in his position for several months and
when he wished to give battle, the officers and troops
dispersed and scattered in all directions. Pu cut his
own throat.*

# 12

When the officers are valiant and the troops ineffective
the army is in distress.[9]

# 13

When senior officers are angry and insubordinate,
and on encountering the enemy rush into battle with
no understanding of the feasibility of engaging and
without awaiting orders from the commander,
the army is in a state of collapse.

---

[7] Wu (伍) denotes a file of five men or the leader of such a file; a corporal;
a non-commissioned officer.

[8] AD 820–5.

[9] Bogged down or sinking, as in a morass. The idea is that if the troops are weak the
efforts of the officers are as vain as if the troops were in a bog.

*Ts'ao Ts'ao:* '*Senior officers' means subordinate generals. If. . . in a rage they attack the enemy without measuring the strength of both sides, then the army is assuredly in a state of collapse.*

# 14

When the general is morally weak and his discipline not strict, when his instructions and guidance are not enlightened, when there are no consistent rules to guide the officers and men and when the formations are slovenly the army is in disorder.[10]

*Chang Yü:. . . Chaos self-induced.*

# 15

When a commander unable to estimate his enemy uses a small force to engage a large one, or weak troops to strike the strong, or when he fails to select shock troops for the van, the result is rout.

*Ts'ao Ts'ao: Under these conditions he commands 'certain-to-flee' troops.*

*Ho Yen-hsi: . . . In the Han the 'Gallants from the Three Rivers' were 'Sword Friends' of unusual talent. In Wu the shock troops were called 'Dissolvers of Difficulty'; in Ch'i 'Fate Deciders'; in the T'ang 'Leapers and Agitators'. These were various names applied to shock troops; nothing is more important in the tactics of winning battles than to employ them.[11]*

*Generally when all the troops are encamped together the general selects from every camp its high-spirited and valiant officers who are distinguished by agility and strength and whose martial accomplishments are above the ordinary. These are grouped to form a special corps. Of ten men, but one is selected; of ten thousand, one thousand.*

*Chang Yü: . . . Generally in battle it is essential to use* élite *troops as the vanguard sharp point. First, because this strengthens our own determination; second, because they blunt the enemy's edge.*

# 16

When any of these six conditions prevails the army is on the road to defeat. It is the highest responsibility of the general that he examine them carefully.

# 17

Conformation of the ground is of the greatest assistance in battle. Therefore, to estimate the enemy situation and to calculate distances and the degree of difficulty of the terrain so as to control victory are virtues of the superior general. He who fights with full knowledge of these factors is certain to win; he who does not will surely be defeated.

# 18

If the situation is one of victory but the sovereign has issued orders not to engage, the general may decide to fight. If the situation is such that he cannot win, but the sovereign has issued orders to engage, he need not do so.

# 19

And therefore the general who in advancing does not seek personal fame, and in withdrawing is not

---

[10] The term rendered 'slovenly' is literally 'vertically and horizontally'.

[11] Unfortunately the functions of the 'Leapers and Agitators' are not explained. Undoubtedly one may have been to arouse the ardour of the troops by wild gyrating and acrobatic sword play for which the Chinese are justly renowned, and possibly at the same time to impress the enemy with their ferocity and skill.

concerned with avoiding punishment, but whose only purpose is to protect the people and promote the best interests of his sovereign, is the precious jewel of the state.

*Li Ch'üan: . . . Such a general has no personal interest.*

*Tu Mu: . . . Few such are to be had.*

# 20

Because such a general regards his men as infants they will march with him into the deepest valleys. He treats them as his own beloved sons and they will die with him.

*Li Ch'üan: If he cherishes his men in this way he will gain their utmost strength. Thus, the Viscount of Ch'u needed but to speak a word and the soldiers felt as if clad in warm silken garments.*[12]

*Tu Mu: During the Warring States when Wu Ch'i was a general he took the same food and wore the same clothes as the lowliest of his troops. On his bed there was no mat; on the march he did not mount his horse; he himself carried his reserve rations. He shared exhaustion and bitter toil with his troops.*

*Chang Yü: . . . Therefore the Military Code says: 'The general must be the first in the toils and fatigues of the army. In the heat of summer he does not spread his parasol nor in the cold of winter don thick clothing. In dangerous places he must dismount and walk. He waits until the army's wells have been dug and only then drinks; until*

---

[12] The Viscount commiserated with those suffering from the cold. His words were enough to comfort the men and raise their flagging spirits.

*the army's food is cooked before he eats; until the army's fortifications have been completed, to shelter himself.'*[13]

# 21

If a general indulges his troops but is unable to employ them; if he loves them but cannot enforce his commands; if the troops are disorderly and he is unable to control them, they may be compared to spoiled children, and are useless.

***Chang Yü:*** *. . . If one uses kindness exclusively the troops become like arrogant children and cannot be employed. This is the reason Ts'ao Ts'ao cut off his own hair and so punished himself.*[14] *. . . Good commanders are both loved and feared. That is all there is to it.*

# 22

If I know that my troops are capable of striking the enemy, but do not know that he is invulnerable to attack, my chance of victory is but half.

# 23

If I know that the enemy is vulnerable to attack, but do not know that my troops are incapable of striking him, my chance of victory is but half.

---

[13] Military essays and codes were generally entitled *Ping Fa*. Chang Yü does not identify the one from which he quotes.

[14] After having issued orders that his troops were not to damage standing grain, Ts'ao Ts'ao carelessly permitted his own grazing horse to trample it. He thereupon ordered himself to be beheaded. His officers tearfully remonstrated, and Ts'ao Ts'ao then inflicted upon himself this symbolic punishment to illustrate that even a commander-in-chief is amenable to military law and discipline.

# 24

If I know that the enemy can be attacked and that my
troops are capable of attacking him, but do not realize
that because of the conformation of the ground I
should not attack, my chance of victory is but half.

# 25

Therefore when those experienced in war move
they make no mistakes; when they act, their resources
are limitless.

# 26

And therefore I say: 'Know the enemy, know yourself;
your victory will never be endangered. Know
the ground, know the weather; your victory
will then be total.'

# THE NINE VARIETIES OF GROUND[1]

Sun Tzu said:

## 1

In respect to the employment of troops, ground may be classified as dispersive, frontier, key, communicating, focal, serious, difficult, encircled, and death.[2]

## 2

When a feudal lord fights in his own territory, he is in dispersive ground.

*Ts'ao Ts'ao: Here officers and men long to return to their nearby homes.*

## 3

When he makes but a shallow penetration into enemy territory he is in frontier ground.[3]

## 4

Ground equally advantageous for the enemy or me to occupy is key ground.[4]

---

[1] The original arrangement of this chapter leaves much to be desired. Many verses are not in proper context; others are repetitious and may possibly be ancient commentary which has worked its way into the text. I have transposed some verses and eliminated those which appear to be accretions. [2] There is some confusion here. The 'accessible' ground of the preceding chapter is defined in the same terms as 'communicating' ground. [3] Lit. 'light' ground, possibly because it is easy to retire or because the officers and men think lightly of deserting just as the expedition is getting under way. [4] This is contestable ground, or, as Tu Mu says, 'strategically important'.

# 5

Ground equally accessible to both the enemy and
me is communicating.
*Tu Mu: This is level and extensive ground in which
one may come and go, sufficient in extent for battle
and to erect opposing fortifications.*

# 6

When a state is enclosed by three other states its
territory is focal. He who first gets control of it will
gain the support of All-under-Heaven.[5]

# 7

When the army has penetrated deep into hostile
territory, leaving far behind many enemy cities and
towns, it is in serious ground.
*Ts'ao Ts'ao: This is ground difficult to return from.*

# 8

When the army traverses mountains, forests,
precipitous country, or marches through defiles,
marshlands, or swamps, or any place where the going
is hard, it is in difficult ground.[6]

# 9

Ground to which access is constricted, where the way
out is tortuous, and where a small enemy force can
strike my larger one is called 'encircled'.[7]
*Tu Mu:. . . Here it is easy to lay ambushes and one
can be utterly defeated.*

---

[5] The Empire is always described as 'All-under-Heaven'.

[6] The commentators indulge in some discussion respecting the interpretation of the
character rendered 'difficult'. Several want to restrict the meaning to ground
susceptible to flooding. [7] The verb may be translated as 'tie down' rather than 'strike'.

# 10

Ground in which the army survives only if it fights
with the courage of desperation is called 'death'.
*Li Ch'üan: Blocked by mountains to the front
and rivers to the rear, with provisions exhausted. In
this situation it is advantageous to act speedily and
dangerous to procrastinate.*

# 11

And therefore, do not fight in dispersive ground;
do not stop in the frontier borderlands.

# 12

Do not attack an enemy who occupies key ground;
in communicating ground do not allow your
formations to become separated.[8]

# 13

In focal ground, ally with neighbouring states;
in deep ground, plunder.[9]

# 14

In difficult ground, press on; in encircled ground,
devise stratagems; in death ground, fight.

# 15

In dispersive ground I would unify the
determination of the army.[10]

---

[8] Ts'ao Ts'ao says they must be 'closed up'.

[9] Li Ch'üan thinks the latter half should read 'do not plunder', as the principal object
when in enemy territory is to win the affection and support of the people.

[10] This and the nine verses which immediately follow have been transposed to this
context. In the text they come later in the chapter.

## 16

In frontier ground I would keep my forces
closely linked.

*Mei Yao-ch'en: On the march the several units are
connected; at halts the camps and fortified posts are
linked together.*

## 17

In key ground I would hasten up my rear elements.

*Ch'ên Hao: What the verse means is that if. . .
the enemy, relying on superior numbers, comes to
contest such ground, I use a large force to hasten into
his rear.*[11]

*Chang Yü:. . . Someone has said that the phrase means 'to set out after the enemy and arrive before him'.*[12]

## 18

In communicating ground I would pay strict attention to my defences.

## 19

In focal ground I would strengthen my alliances.
*Chang Yü: I reward my prospective allies with valuables and silks and bind them with solemn covenants. I abide firmly by the treaties and then my allies will certainly aid me.*

## 20

In serious ground I would ensure a continuous flow of provisions.

## 21

In difficult ground I would press on over the roads.

## 22

In encircled ground I would block the points of access and egress.
*Tu Mu: It is military doctrine that an encircling force must leave a gap to show the surrounded troops there is a way out, so that they will not be determined to fight to the death. Then, taking advantage of this, strike. Now, if I am in encircled*

---

[11] The question is, whose 'rear' is Sun Tzu talking about? Ch'ên Hao is reading something into the verse as it stands in present context.

[12] The 'someone' is Mei Yao-ch'en, who takes *hou* (後) to mean 'after' in the temporal sense.

*ground, and the enemy opens a road in order to tempt my troops to take it, I close this means of escape so that my officers and men will have a mind to fight to the death.*[13]

## 23

In death ground I could make it evident that there is no chance of survival. For it is the nature of soldiers to resist when surrounded; to fight to the death when there is no alternative, and when desperate to follow commands implicitly.

## 24

The tactical variations appropriate to the nine types of ground, the advantages of close or extended deployment, and the principles of human nature are matters the general must examine with the greatest care.[14]

## 25

Anciently, those described as skilled in war made it impossible for the enemy to unite his van and his rear; for his elements both large and small to mutually co-operate; for the good troops to succour the poor and for superiors and subordinates to support each other.[15]

---

[13] A long story relates that Shen Wu of the Later Wei, when in such a position, blocked the only escape road for his troops with the army's livestock. His forces then fought desperately and defeated an army of two hundred thousand.

[14] This verse is followed by seven short verses which again define terms previously defined in v. 2 to 10 inclusive. This appears to be commentary which has worked its way into the text.

[15] The implication is that even were the enemy able to concentrate, internal dissensions provoked by the skilled general would render him ineffective.

# 26

When the enemy's forces were dispersed they prevented him from assembling them; when concentrated, they threw him into confusion.

*Meng: Lay on many deceptive operations. Be seen in the west and march out of the east; lure him in the north and strike in the south. Drive him crazy and bewilder him so that he disperses his forces in confusion.*

*Chang Yü: Take him unaware by surprise attacks where he is unprepared. Hit him suddenly with shock troops.*

# 27

They concentrated and moved when it was advantageous to do so;[16] when not advantageous, they halted.

# 28

Should one ask: 'How do I cope with a well ordered enemy host about to attack me?' I reply: 'Seize something he cherishes and he will conform to your desires.'[17]

# 29

Speed is the essence of war. Take advantage of the enemy's unpreparedness; travel by unexpected routes and strike him where he has taken no precautions.

*Tu Mu: This summarizes the essential nature of war. . . and the ultimate in generalship.*

迅
速
的

---

[16] Lit. 'They concentrated where it was advantageous to do so and then acted. When it was not advantageous, they stood fast.' In another commentary Shih Tzu-mei says not to move unless there is advantage in it.

[17] Comments between question and answer omitted.

***Chang Yü:*** *Here Sun Tzu again explains. . . that the one thing esteemed is divine swiftness.*

# 30

The general principles applicable to an invading force are that when you have penetrated deeply into hostile territory your army is united, and the defender cannot overcome you.

# 31

Plunder fertile country to supply the army with plentiful provisions.

# 32

Pay heed to nourishing the troops; do not unnecessarily fatigue them. Unite them in spirit; conserve their strength. Make unfathomable plans for the movements of the army.

# 33

Throw the troops into a position from which there is no escape and even when faced with death they will not flee. For if prepared to die, what can they not achieve? Then officers and men together put forth their utmost efforts. In a desperate situation they fear nothing; when there is no way out they stand firm. Deep in a hostile land they are bound together, and there, where there is no alternative, they will engage the enemy in hand to hand combat.[18]

---

[18] There are several characters in Chinese which basically mean 'to fight'. That used here implies 'close combat'.

[19] This refers to the troops of a general who nourishes them, who unites them in spirit, who husbands their strength, and who makes unfathomable plans.

# 34

Thus, such troops need no encouragement
to be vigilant. Without extorting their support
the general obtains it; without inviting their
affection he gains it; without demanding their
trust he wins it.[19]

# 35

My officers have no surplus of wealth but not because
they disdain worldly goods; they have no expectation
of long life but not because they dislike longevity.
*Wang Hsi:. . . When officers and men care only for
worldly riches they will cherish life at all costs.*

# 36

On the day the army is ordered to march the tears
of those seated soak their lapels; the tears of those
reclining course down their cheeks.
*Tu Mu: All have made a covenant with death.
Before the day of battle the order is issued:
'Today's affair depends upon this one stroke. The
bodies of those who do not put their lives at stake
will fertilize the fields and become carrion for the
birds and beasts.'*

# 37

But throw them into a situation where there is no
escape and they will display the immortal courage
of Chuan Chu and Ts'ao Kuei.[20]

---

[20] The exploits of these heroes are recounted in SC, ch. 68.

# 38

Now the troops of those adept in war are used like the 'Simultaneously Responding' snake of Mount Ch'ang. When struck on the head its tail attacks; when struck on the tail, its head attacks, when struck in the centre both head and tail attack.[21]

# 39

Should one ask: 'Can troops be made capable of such instantaneous co-ordination?' I reply: 'They can.' For, although the men of Wu and Yüeh mutually hate one another, if together in a boat tossed by the wind they would co-operate as the right hand does with the left.'

# 40

It is thus not sufficient to place one's reliance on hobbled horses or buried chariot wheels.[22]

# 41

To cultivate a uniform level of valour is the object of military administration.[23] And it is by proper use of the ground that both shock and flexible forces are used to the best advantage.[24]

*Chang Yü: If one gains the advantage of the ground then even weak and soft troops can conquer the enemy. How much more so if they are tough and strong! That both may be used effectively is because they are disposed in accordance with the conditions of the ground.*

# 42

It is the business of a general to be serene and inscrutable, impartial and self-controlled,[25]

*Wang Hsi: If serene he is not vexed; if inscrutable, unfathomable; if upright, not improper; if self-controlled, not confused.*

---

[21] This mountain was anciently known as Mt. Hêng. During the reign of the Emperor Wên (Liu Hêng) of the Han (179–159 BC) the name was changed to 'Ch'ang' to avoid the taboo. In all existing works 'Hêng' was changed to 'Ch'ang'.

[22] Such 'Maginot Line' expedients are not in themselves sufficient to prevent defending troops from fleeing. [23] Lit. 'To equalize courage [so that it is that of] one [man] is the right way of administration.'

[24] Chang Yü makes it clear why terrain should be taken into account when troops are disposed. The difference in quality of troops can be balanced by careful sector assignment. Weak troops can hold strong ground, but might break if posted in a position less strong.

[25] Giles translated: 'It is the business of a general to be quiet and thus ensure secrecy; upright and just and thus maintain order.' The commentators do not agree, but none takes it in this sense, nor does the text support this rendering. I follow Ts'ao Ts'ao and Wang Hsi.

# 43

He should be capable of keeping his officers and men
in ignorance of his plans.
*Ts'ao Ts'ao:. . . His troops may join him in rejoicing
at the accomplishment, but they cannot join him in
laying the plans.*

# 44

He prohibits superstitious practices and so rids the
army of doubts. Then until the moment of death there
can be no troubles.[26]
*Ts'ao Ts'ao: Prohibit talk of omens and of supernat-
ural portents. Rid plans of doubts and uncertainties.*
*Chang Yü: The* Ssu-ma Fa *says: 'Exterminate
superstitions.'*

# 45

He changes his methods and alters his plans so that
people have no knowledge of what he is doing.
*Chang Yü: Courses of action previously followed
and old plans previously executed must be altered.*

# 46

He alters his camp-sites and marches by devious routes,
and thus makes it impossible for others to anticipate
his purpose.[27]

---

[26] The 之 at the end of this sentence is emended to read 災 which means a
natural or 'heaven sent' calamity. Part of Ts'ao Ts'ao's comment which is omitted
indicates that various texts were circulating in his time.

[27] Or perhaps, 'makes it impossible for the enemy to learn his plans'. But Mei Yao-
ch'en thinks the meaning is that the enemy will thus be rendered incapable of laying
plans. Giles infers that the general, by altering his camps and marching by devious
routes, can prevent the enemy 'from anticipating his purpose', which seems the best.
The comments do not illuminate the point at issue.

# 47

To assemble the army and throw it into a desperate
position is the business of the general.

# 48

He leads the army deep into hostile territory and there
releases the trigger.[28]

# 49

He burns his boats and smashes his cooking pots;
he urges the army on as if driving a flock of sheep,
now in one direction, now in another, and none knows
where he is going.[29]

# 50

He fixes a date for rendezvous and after the troops have
met, cuts off their return route just as if he were
removing a ladder from beneath them.

# 51

One ignorant of the plans of neighbouring
states cannot prepare alliances in good time; if
ignorant of the conditions of mountains, forests,
dangerous defiles, swamps and marshes he cannot
conduct the march of an army; if he fails to make use
of native guides he cannot gain the advantages of the

---

[28] 'Release' of a trigger, or mechanism, is the usual meaning of the expression
(發 機). The idiom has been translated: 'puts into effect his expedient plans.'
Wang Hsi says that when the trigger is released 'there is no return' (of the arrow
or bolt). Lit. this verse reads: 'He leads the army deep into the territory of the
Feudal Lords and there releases the trigger' (or 'puts into effect his expedient plans').
Giles translates the phrase in question as 'shows his hand',
i.e. takes irrevocable action.
[29] Neither his own troops nor the enemy can fathom his ultimate design.

ground. A general ignorant of even one of these three matters is unfit to command the armies of a Hegemonic King.[30]

*Ts'ao Ts'ao: These three matters have previously been elaborated. The reason Sun Tzu returns to the subject is that he strongly disapproved of those unable to employ troops properly.*

## 52

Now when a Hegemonic King attacks a powerful state he makes it impossible for the enemy to concentrate. He overawes the enemy and prevents his allies from joining him.[31]

*Mei Yao-ch'en: In attacking a great state, if you can divide your enemy's forces your strength will be more than sufficient.*

## 53

It follows that he does not contend against powerful combinations nor does he foster the power of other states. He relies for the attainment of his aims on his ability to overawe his opponents. And so he can take the enemy's cities and overthrow the enemy's state.[32]

*Ts'ao Ts'ao: By 'Hegemonic King' is meant one who*

---

[30] Emending 四 五 者—'[these] four or five [matters]'—to read 此 三 者— 'these three [matters]'

[31] This verse and the next present problems. Chang Yü thinks the verse means that if the troops of a Hegemonic King (or a ruler who aspires to such status) attack hastily (or recklessly, or without forethought) his allies will not come to his aid. The other commentators interpret the verse as I have.

[32] The commentators differ in their interpretations of this verse. Giles translates: 'Hence he does not strive to ally himself with all and sundry nor does he foster the power of other states. He carries out his own secret designs, keeping his antagonists in awe. Thus he is able to capture their cities and overthrow their kingdoms.' But I feel that Sun Tzu meant that the 'Hegemonic King' need not contend against

*does not ally with the feudal lords. He breaks up the alliances of All-under-Heaven and snatches the position of authority. He uses prestige and virtue to attain his ends.*[33]

***Tu Mu:*** *The verse says if one neither covenants for the help of neighbours nor develops plans based on expediency but in furtherance of his personal aims relies only on his own military strength to overawe the enemy country then his own cities can be captured and his own state overthrown.*[34]

---

'powerful combinations' because he isolates his enemies. He does not permit them to form 'powerful combinations'.

[33] Possibly Giles derived his interpretation from this comment.

[34] Also a justifiable interpretation, which illustrates how radically the commentators frequently differ.

# 54

Bestow rewards without respect to customary practice;
publish orders without respect to precedent.[35] Thus you
may employ the entire army as you would one man.
*Chang Yü: . . . If the code respecting rewards and
punishments is clear and speedily applied then you
may use the many as you do the few.*

# 55

Set the troops to their tasks without imparting your
designs; use them to gain advantage without revealing
the dangers involved. Throw them into a perilous
situation and they survive; put them in death ground
and they will live. For when the army is placed in
such a situation it can snatch victory from defeat.

# 56

Now the crux of military operations lies in the
pretence of accommodating one's self to the
designs of the enemy.[36]

# 57

Concentrate your forces against the enemy and from
a distance of a thousand *li* you can kill his general.[37]
This is described as the ability to attain one's aim
in an artful and ingenious manner.

---

[35] This verse, obviously out of place, emphasizes that the general in the field need
not follow prescribed procedures in recognition of meritorious service but should
bestow timely rewards. The general need not follow customary law in respect to
administration of his army.

[36] Possibly too free a translation, but the commentators agree that this is the idea
Sun Tzu tries to convey. I follow Tu Mu.

[37] I follow Ts'ao Ts'ao here. A strategist worthy of the name defeats his enemy from a
distance of one thousand *li* by anticipating his enemy's plans.

# 58

On the day the policy to attack is put into effect, close the passes, rescind the passports,[38] have no further intercourse with the enemy's envoys and exhort the temple council to execute the plans.[39]

# 59

When the enemy presents an opportunity, speedily take advantage of it.[40] Anticipate him in seizing something he values and move in accordance with a date secretly fixed.

# 60

The doctrine of war is to follow the enemy situation in order to decide on battle.[41]

# 61

Therefore at first be shy as a maiden. When the enemy gives you an opening be swift as a hare and he will be unable to withstand you.

---

[38] Lit. 'break the tallies'. These were carried by travellers and were examined by the Wardens of the Passes. Without a proper tally no one could legally enter or leave a country.

[39] The text is confusing. It seems literally to read: 'From [the rostrum of] temple, exhort [the army?] [the people?] to execute the plans.' The commentators are no help.

[40] Another difficult verse. Some commentators think it should read: 'When the enemy sends spies, immediately let them enter.' The difficulty is in the idiom k'ai ho (開 闔), literally, 'to open the leaf of a door', thus, 'to present an opportunity [to enter]'. Ts'ao Ts'ao says the idiom means 'a cleavage', 'a gap', or 'a space'. Then, he goes on, 'you must speedily enter'. Other commentators say the idiom means 'spies' or 'secret agents'. I follow Ts'ao Ts'ao.

[41] The commentators again disagree: v. 58–61 are susceptible to varying translations or interpretations.

# ATTACK BY FIRE

Sun Tzu said:

## 1

There are five methods of attacking with fire. The first is to burn personnel; the second, to burn stores; the third, to burn equipment; the fourth, to burn arsenals; and the fifth, to use incendiary missiles.[1]

## 2

To use fire, some medium must be relied upon. *Ts'ao Ts'ao: Rely upon traitors among the enemy.[2] Chang Yü: All fire attacks depend on weather conditions.*

## 3

Equipment for setting fires must always be at hand. *Chang Yü: Implements and combustible materials should be prepared beforehand.*

## 4

There are suitable times and appropriate days on which to raise fires.

---

[1] There is a mistake in the text here. Tu Yu emends and explains that flame-tipped arrows are fired into the enemy's barracks or camp by strong crossbowmen. Other commentators vary in their interpretations, but Tu Yu's emendation is logical.

[2] 'among the enemy' added. Ch'ên Hao remarks that one does not only rely on traitors.

# 5

'Times' means when the weather is scorching hot;
'days' means when the moon is in Sagittarius, Alpharatz,
*I,* or *Chen* constellations, for these are days of rising winds.[3]

# 6

Now in fire attacks one must respond to the
changing situation.

# 7

When fire breaks out in the enemy's camp immediately
co-ordinate your action from without. But if his troops
remain calm bide your time and do not attack.

# 8

When the fire reaches its height, follow up if you can.
If you cannot do so, wait.

# 9

If you can raise fires outside the enemy camp, it is not necessary to wait until they are started inside. Set fires at suitable times.[4]

# 10

When fires are raised up-wind do not attack from down-wind.

# 11

When the wind blows during the day it will die down at night.[5]

---

[3] Sun Hsing-yen has emended the original text in accordance with the TT and YL, but the original seems better and I follow it. I cannot place the *I* and *Chen* constellations.

[4] A warning not to be cooked in your own fire is to be inferred from the last sentence.

[5] Following Chang Yu.

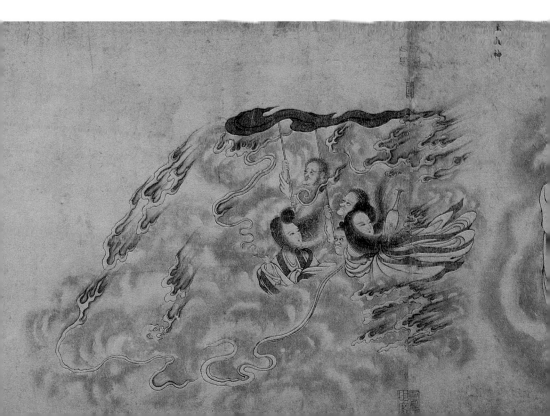

## 12

Now the army must know the five different fire-attack
situations and be constantly vigilant.[6]

## 13

Those who use fire to assist their attacks are intelligent;
those who use inundations are powerful.

## 14

Water can isolate an enemy but cannot destroy his
supplies or equipment.[7]

# 15

Now to win battles and take your objectives, but to fail to exploit these achievements is ominous and may be described as 'wasteful delay'.[8]

# 16

And therefore it is said that enlightened rulers deliberate upon the plans, and good generals execute them.

# 17

If not in the interests of the state, do not act. If you cannot succeed, do not use troops. If you are not in danger, do not fight.[9]

# 18

A sovereign cannot raise an army because he is enraged, nor can a general fight because he is resentful. For while an angered man may again be happy, and a resentful man again be pleased, a state that has perished cannot be restored, nor can the dead be brought back to life.

# 19

Therefore, the enlightened ruler is prudent and the good general is warned against rash action.[10] Thus the state is kept secure and the army preserved.

---

[6] Following Tu Mu. [7] Following Ts'ao Ts'ao.

[8] Mei Yao-ch'en is the only commentator who caught Sun Tzu's meaning. situations must be exploited.

[9] The commentators make it clear that war is to be used only as a last resort.

[10] Last three words added. Rage and resentment lead to rash action.

# EMPLOYMENT OF SECRET AGENTS[1]

Sun Tzu said:

## 1

Now when an army of one hundred thousand is raised and dispatched on a distant campaign the expenses borne by the people together with the disbursements of the treasury will amount to a thousand pieces of gold daily. There will be continuous commotion both at home and abroad, people will be exhausted by the requirements of transport, and the affairs of seven hundred thousand households will be disrupted.[2]

*Tiao Tsao: Anciently, eight families comprised a community. When one family sent a man to the army, the remaining seven contributed to its support. Thus, when an army of one hundred thousand was raised those unable to attend fully to their own ploughing and sowing amounted to seven hundred thousand households.*

## 2

One who confronts his enemy for many years in order to struggle for victory in a decisive battle yet who,

---

[1] The character used in the title means 'the space between' two objects (such as a crack between two doors) and thus 'cleavage', 'division', or 'to divide', It also means 'spies', 'spying', or 'espionage'.

[2] I have translated 'to a distance of one thousand *li*' as 'on a distant campaign'. The figure need not be taken as specific.

because he begrudges rank, honours and a few
hundred pieces of gold, remains ignorant of his
enemy's situation, is completely devoid of humanity.
Such a man is no general; no support to his sovereign;
no master of victory.

## 3

Now the reason the enlightened prince and the
wise general conquer the enemy whenever they move
and their achievements surpass those of ordinary men
is foreknowledge.

**Ho Yen-hsi:** *The section in the Rites of Chou entitled 'Military Officers' names 'The Director of
National Espionage'. This officer probably directed
secret operations in other countries.*[3]

## 4

What is called 'foreknowledge' cannot be elicited from
spirits, nor from gods, nor by analogy with past events,
nor from calculations. It must be obtained from men
who know the enemy situation.

## 5

Now there are five sorts of secret agents to be
employed. These are native, inside, doubled,
expendable, and living.[4]

## 6

When these five types of agents are all working
simultaneously and none knows their method of

---

[3] Probably an appeal to the authority of tradition to support the legitimacy
of espionage and subversion which are contrary to the spirit of
Confucian teaching.

[4] I use 'expendable' in lieu of 'death'.

operation, they are called 'The Divine Skein' and are the treasure of a sovereign.[5]

7

Native agents are those of the enemy's country people whom we employ.

8

Inside agents are enemy officials whom we employ.

*Tu Mu: Among the official class there are worthy men who have been deprived of office; others who have committed errors and have been punished. There are sycophants and minions who are covetous of wealth. There are those who wrongly remain long in lowly office; those who have not obtained responsible positions, and those whose sole desire is to take advantage of times of trouble to extend the scope of their own abilities. There are those who are two-faced, changeable, and deceitful, and who are always sitting on the fence. As far as all such are concerned you can secretly inquire after their welfare, reward them liberally with gold and silk, and so tie them to you. Then you may rely on them to seek out the real facts of the situation in their country, and to ascertain its plans directed against you. They can as well create cleavages between the sovereign and his ministers so that these are not in harmonious accord.*

9

Doubled agents are enemy spies whom we employ.

*Li Ch'üan. When the enemy sends spies to pry*

秘
密
的

---

[5] The idea is that information may be gathered in as fish are by pulling on a single cord and so drawing together the various threads of a net.

*into my accomplishments or lack of them, I bribe them lavishly, turn them around, and make them my agents.*

## 10

Expendable agents are those of our own spies who are deliberately given fabricated information.

*Tu Yu: We leak information which is actually false and allow our own agents to learn it. When these agents operating in enemy territory are taken by him they are certain to report this false information. The*

*enemy will believe it and make preparations accordingly. But our actions will of course not accord with this, and the enemy will put the spies to death.*

*Chang Yü: . . . In our dynasty Chief of Staff Ts'ao once pardoned a condemned man whom he then disguised as a monk, and caused to swallow a ball of wax and enter Tangut. When the false monk arrived he was imprisoned. The monk told his captors about the ball of wax and soon discharged it in a stool. When the ball was opened, the Tanguts read a letter transmitted by Chief of Staff Ts'ao to their Director of Strategic Planning. The chieftain of the barbarians was enraged, put his minister to death, and executed the spy monk. This is the idea. But expendable agents are not confined to only one use. Sometimes I send agents to the enemy to make a covenant of peace and then I attack.*

# 11

Living agents are those who return with information.

*Tu Yu: We select men who are clever, talented, wise, and able to gain access to those of the enemy who are intimate with the sovereign and members of the nobility. Thus they are able to observe the enemy's movements and to learn of his doings and his plans. Having learned the true state of affairs they return and tell us. Therefore they are called 'living' agents.*

*Tu Mu: These are people who can come and go and communicate reports. As living spies we must recruit men who are intelligent but appear to be stupid; who seem to be dull but are strong in heart; men who are agile, vigorous, hardy, and brave; well-versed in lowly matters and able to endure hunger, cold, filth, and humiliation.*

## 12

Of all those in the army close to the commander none is more intimate than the secret agent; of all rewards none more liberal than those given to secret agents; of all matters none is more confidential than those relating to secret operations.

*Mei Yao-ch'en: Secret agents receive their instructions within the tent of the general, and are intimate and close to him.*

*Tu Mu: These are 'mouth to ear' matters.*

## 13

He who is not sage and wise, humane and just, cannot use secret agents. And he who is not delicate and subtle cannot get the truth out of them.

*Tu Mu: The first essential is to estimate the character of the spy to determine if he is sincere, truthful, and really intelligent. Afterwards, he can be employed. . . . Among agents there are some whose only interest is in acquiring wealth without obtaining the true situation of the enemy, and only meet my requirements with empty words.[6] In such a case I must be deep and subtle. Then I can assess the truth or falsity of the spy's statements and discriminate between what is substantial and what is not.*

*Mei Yao-ch'en: Take precautions against the spy having been turned around.*

## 14

Delicate indeed! Truly delicate! There is no place where espionage is not used.

---

[6] Such agents are now aptly described as 'paper mills'.

# 15

If plans relating to secret operations are prematurely divulged the agent and all those to whom he spoke of them shall be put to death.[7]

*Ch'en Hao: . . . They may be killed in order to stop their mouths and prevent the enemy hearing.*

# 16

Generally in the case of armies you wish to strike, cities you wish to attack, and people you wish to assassinate, you must know the names of the garrison commander, the staff officers, the ushers, gate keepers, and the bodyguards. You must instruct your agents to inquire into these matters in minute detail.

*Tu Mu: If you wish to conduct offensive war you must know the men employed by the enemy. Are they wise or stupid, clever or clumsy? Having assessed their qualities, you prepare appropriate measures. When the King of Han sent Han Hsin, Ts'ao Ts'an, and Kuan Ying to attack Wei Pao he asked: 'Who is the commander-in-chief of Wei?' The reply was: 'Po Chih.' The King said: 'His mouth still smells of his mother's milk. He cannot equal Han Hsin. Who is the cavalry commander?' The reply was: 'Feng Ching.' The King said: 'He is the son of General Feng Wu-che of Ch'in. Although worthy, he is not the equal of Kuan Ying. And who is the infantry commander?' The reply was: 'Hsiang T'o.' The King said: 'He is no match for Ts'ao Ts'an. I have nothing to worry about.'*

---

[7] Giles translated: 'If a secret piece of news is divulged by a spy before the time is ripe. . . .' Sun Tzu is not talking about 'news' here but about espionage affairs, or matters or plans relating to espionage.

# 17

It is essential to seek out enemy agents who have come
to conduct espionage against you and to bribe them to
serve you. Give them instructions and care for them.[8]
Thus doubled agents are recruited and used.

# 18

It is by means of the doubled agent that native and
inside agents can be recruited and employed.
*Chang Yü: This is because the doubled agent knows
those of his own countrymen who are covetous as
well as those officials who have been remiss in office.
These we can tempt into our service.*

# 19

And it is by this means that the expendable agent,
armed with false information, can be sent to convey it
to the enemy.
*Chang Yü: It is because doubled agents know
in what respects the enemy can be deceived that
expendable agents may be sent to convey false
information.*

# 20

It is by this means also that living agents can be used at
appropriate times.

---

[8] These agents, according to Giles' translation, are to be 'tempted with bribes, led
away and comfortably housed'.

# 21

The sovereign must have full knowledge of the activities of the five sorts of agents. This knowledge must come from the doubled agents, and therefore it is mandatory that they be treated with the utmost liberality.

秘
密
的

# 22

Of old, the rise of Yin was due to I Chih, who formerly served the Hsia; the Chou came to power through Lu Yu, a servant of the Yin.[9]
*Chang Yü: I Chih was a minister of Hsia who went over to the Yin. Lu Wang was a minister of Yin who went over to the Chou.*

# 23

And therefore only the enlightened sovereign and the worthy general who are able to use the most intelligent people as agents are certain to achieve great things. Secret operations are essential in war; upon them the army relies to make its every move.
*Chia Lin: An army without secret agents is exactly like a man without eyes or ears.*

---

[9] Several of the commentators are outraged that these worthies are described by Sun Tzu as 'spies' or 'agents', but of course they were

# WU CH'I'S
# 'ART OF WAR'

## A NOTE ON WU CH'I

Wu Ch'i, whose name is always associated with Sun Tzu's, was born in Wei about 430 BC and executed in Ch'u in 381 BC. In his youth he was a pupil of Tsêng Ts'an, who conceived a dislike for him and banished him from his presence. Proceeding to Lu, Wu Ch'i studied the art of war, and soon became recognized as an expert. When hostilities broke out between Lu and Ch'i, he was anxious to take command of the Lu army, but the Prince hesitated to appoint him because his wife was a native of Ch'i. Wu Ch'i thereupon murdered her to demonstrate his loyalty, and entered upon what proved to be a successful military career. Later he took service with the Wei State, where for some time he enjoyed the favour of Marquis Wu. On one occasion, while navigating the West River, the Marquis remarked upon the splendid natural defences of that region, to which Wu Ch'i replied that the virtue of its ruler is a far greater safeguard to a state than a frontier of inaccessible cliffs. In 387 BC, having fallen into disfavour and believing his life to be in danger, he fled to Ch'u, where he became Chancellor. Here he reorganized the administration. By the unsparing severity with which he abolished all abuses, he made himself many foes among the chief families. After the death of his patron, King Tao, he was killed. As a general he was severe, but gained the affections of his troops by sharing every hardship with them. The work ascribed to him is undoubtedly a compilation prepared after his death.

## CHAPTER I

PLANNING OPERATIONS
AGAINST OTHER STATES

### SECTION I

1. Wu Ch'i, clad in a Confucian robe, made use of his knowledge of military affairs to gain audience with Marquis Wen of Wei.

2. Marquis Wen said: 'I do not care for military affairs.'

Wu Ch'i said: 'I am able to inquire into what is hidden and by means of the past investigate the course of future events. Why are your words, My Lord, so different from your thoughts?

3. 'At present, My Lord, during the four seasons you cause animals to be skinned and lacquer their hides and paint them vermilion and blue.

You brilliantly decorate them with rhinoceros horn and ivory.

'If you wear these in the winter you are not warm, and in the summer, not cool. You make spears twenty-four feet long, and short halberds of half this length. You cover the wheels and doors of your chariots with leather; they are not pleasing to the eyes, and when used for hunting they are not light.

'I do not comprehend how you, My Lord, propose to use them.

4. 'If these are made ready for offensive or defensive war and you do not seek men able to use such equipment it would be like chickens fighting a fox, or puppies which attack a tiger. Though they have fighting hearts, they will perish.

5. 'Anciently the Lord of the Ch'eng Shang tribe paid his entire attention to cultivating his virtue and eschewed military affairs. As a result his state

was extinguished. There was the Lord of Yu Hu who placed his entire reliance in the number and valour of his troops and thereby lost his altars of Land and Grain.

'The enlightened ruler who takes warning from these precedents would assuredly in the capital promote learning and virtue, and in the field prepare for defence. Therefore, a ruler who is unable to advance when he confronts the enemy is not Righteous and one who looks upon the corpses of those killed in battle and mourns them is not Benevolent.'

6. Whereupon Duke Wen personally spread the mats and his wife respectfully offered Wu Ch'i a goblet of wine. Sacrifices were made in the ancestral temple and Wu Ch'i was named Commander-in-Chief. He guarded the western rivers and fought seventy-six battles with the Feudal Lords. Of these he gained complete victory in sixty-four, while the remaining were drawn. He opened up new lands in every direction and extended the boundaries for a thousand *li*. All these were the achievements of Wu Ch'i.

### SECTION II

Wu Tzu said:

1. Anciently, those who planned against another state would surely first instruct the hundred clans and then display affection toward the myriad people.

2. There are four matters in which concord may be lacking. When there is discord within the country the army cannot be mobilized. When there is discord in the army it cannot take the field. When there is lack of harmony in the field the army cannot take the offensive. When there is lack of harmony in battle the army cannot

win a decisive victory. Therefore the generals of a sovereign who follows 'The Right Way', when about to employ the people, first establish concord and then undertake matters of great importance.

3. Such a sovereign does not presume to place confidence in his own personal plans, but necessarily discusses them in the Temple of the Ancestors after tortoise divination and reflection upon the suitable season. If the omens are auspicious he then launches the army.

4. When the people know that the sovereign loves their lives and sorrows at their deaths to such an extent that he will face crisis together with them, the officers will consider it glorious to advance and die and shameful to save their lives by retreat.

### SECTION III

Wu Tzu said:

1. Now what is called 'The Right Way' is the return to fundamental principles; 'Righteousness' is that by which affairs are advanced and merit established; 'Planning' is that by which harm is avoided and advantage gained; 'Essentials' that which safeguards one's work and protect his achievements. If conduct is not in accord with 'The Right Way' and action not in accord with Righteousness, then albeit one's position is important and honourable, misfortune will overtake him.

2. And therefore surely the Sage controls the people with the highest principles and governs them with Righteousness. He stimulates them with ritual and soothes them with humane treatment. When these four Virtues are practised the people flourish; when they are neglected the people decline. Therefore when T'ang the

Victorious attacked Chieh the people of Hsia rejoiced; when Wu of Chou attacked Chou Hsin of the Shang dynasty the people of Yin did not oppose him. They acted in accord with the will of Heaven and that of Man, and thus were able to achieve these things.

### SECTION IV

Wu Tzu said:

1. Generally in administering a country and controlling an army it is necessary to instruct the people by using Ritual and to encourage them with Righteousness so as to inculcate the sense of honour. Now if men's sense of honour is great they will be able to campaign; if less, they will be able to defend. To win victory is easy; to preserve its fruits, difficult. And therefore it is said that when All-under-Heaven is at war, one who gains five victories suffers calamity, one who gains four is exhausted; one who gains three becomes Lord Protector; one who gains two, a King; one who gains one, the Emperor. Thus he who by countless victories has gained empire is unique, while those who have perished thereby are many.

### SECTION V

Wu Tzu said:

1. Now there are five matters which give rise to military operations. First, the struggle for fame; second, the struggle for advantage; third, the accumulation of animosity; fourth, internal disorder; and fifth, famine.

2. There are also five categories of war. First, righteous war; second, aggressive war; third, enraged war; fourth, wanton war; and fifth insurgent war. Wars to suppress violence and quell disorder are Righteous. Those which

depend on force are aggressive. When troops are raised because rulers are actuated by anger, this is enraged war. Those in which all propriety is discarded because of greed are wanton wars. Those who, when the state is in disorder and the people exhausted, stir up trouble and agitate the multitude, cause insurgent wars.

3. There is a suitable method for dealing with each: a righteous war must be forestalled by proper government; an aggressive war by humbling one's self; an enraged war by reason; a wanton war by deception and treachery; and an insurgent war by authority.

4. Marquis Wu asked: 'I wish to know the way to control troops, to estimate my enemy and to strengthen the state.'

Wu Ch'i replied: 'Of old, enlightened kings assuredly paid attention to the appropriate relationship of prince and minister and of superiors and subordinates. They peacefully gathered together officials and people and instructed them according to custom. They selected and summoned worthy men of talent, in order to prepare themselves for any contingency.

5. 'Anciently, Huan of Ch'i levied fifty thousand warriors and became Lord Protector of the feudal states. Wen of Chin called up forty thousand men for the van army and achieved his ambitions. Mu of Ch'in with thirty thousand valiant troops subdued his neighbouring enemies. Therefore the ruler of a powerful state must estimate his own people. Those who are bold, spirited, and strong he will form into one detachment. Those who rejoice at advancing into battle and exert themselves to demonstrate their loyalty and valour he will form into one detachment. Those able to climb heights and leap far, who are nimble and fleet of foot, he will form into one detachment. Princes and ministers who have lost position and who wish to acquire merit in the eyes of their superiors he will form into one detachment. Those who abandoned the cities they were defending and who wish to rectify their shameful conduct, he will gather into one detachment. 'These five will form a well-trained and keen army. If you have three thousand men like this they can break an encirclement from within, or from without can slaughter the defenders of the enemy's cities.'

## SECTION VI

1. Marquis Wu asked: 'I should like to know the way to make my battle formations certainly firm, my defence strong, and how in battle to be certain of winning.'

Wu Ch'i replied: 'These things may be seen at once. How is it that you wish to hear of them? If Your Majesty can employ the worthy in high position, and those who are worthless in inferior position, then the array will be already firm. If people are secure in their farms and dwellings and friendly with their magistrates, then your defences are already strong. If the clans approve of their own sovereign and disapprove of others, then the battles are already won.'

2. The Marquis of Wu was once deliberating on state affairs and none of his ministers' opinions was equal to his. He retired from the court looking pleased. Wu Ch'i advanced and said: 'Anciently, King Chuang of Ch'u was deliberating on state affairs. None of his ministers' opinions could equal his. He retired from the council looking worried. Lord Shen asked: "Why does the Sovereign look worried?" The King replied: "This humble one has heard that the world never lacks

sages and that a country never lacks wise men. One able to get a sage for his teacher will be a King; one able to get a wise man for his friend, a Lord Protector. Now I have no talent, and still my ministers cannot equal me. Ch'u is in danger.' This is what worried King Chuang of Ch'u, but pleases you. I, your servant, am secretly apprehensive.' Whereupon Marquis Wu looked ashamed.

## CHAPTER II

### ESTIMATING THE ENEMY

#### SECTION I

1. Marquis Wu said to Wu Ch'i: 'Now Ch'in intimidates me from the west; Ch'u girdles me on the south; Chao confronts me in the north; Ch'i overlooks my eastern borders, Yen cuts off my rear and Han takes position to my front. I must defend on all four sides against the troops of these six states. This situation is most inconvenient. How am I to deal with these worries!'

Wu Ch'i replied: 'Now the way to make the country secure is to prize precaution. Now that you are aware of the dangers, misfortune is kept at a distance.

2. 'I beg to discuss the customs of these six countries. Now the Ch'i army is massive but not firm. The Ch'in army is dispersed and each fights for himself. The Ch'u army is well organized but has no endurance. The Yen army will defend, but not take the field. The armies of the Three Chins are well administered but not used. Now the men of Ch'i are hardy, the state is rich, the sovereign and ministers are arrogant and extravagant and treat the people with contempt. The government is lenient but emoluments are inequitable. Its army is of two minds, and its weight is at the front, not at the rear. Therefore although massive it is not firm. The way to attack the Ch'i army is this. It must be divided into three parts and its right and left attacked. So you force them to conform to you and their army can be destroyed.

'The character of the Ch'in State is strong; its country precipitous; its government strict. Its rewards and punishments are to be trusted. Its people are unyielding and all individuals are determined to fight. Therefore its formations are dispersed and each fights for itself. The way to attack Ch'in is first to offer them some apparent advantage and entice them by retiring. The officers will covet the bait and will become separated from the generals. Take advantage of their errors and hunt down their dispersed elements. Prepare ambushes, seize opportunities, and their generals can be taken.

'The character of the Ch'u State is weak; its territory extensive. Its government is vacillating and its people exhausted. Therefore although well organized it is not enduring. The way to attack Ch'u is to strike suddenly and throw their camps into confusion. First, deprive them of morale. Advance with light troops, withdraw speedily, thus exhausting and tiring them. Without contesting with them in battle, their armies can be defeated.

'The character of the Yen people is that they are stupid and honest. They are cautious. They admire valour and righteousness but are lacking in trickery and deception. Therefore they will defend but will not take the field. The way to attack Yen is to stir them up and harass them, advance upon them and then withdraw to a distance. Swiftly get into their rear. Then the superiors will be perplexed and the subordinates fearful. They will take precautions against our chariots and cavalry and will fall back before them. And thus their generals can be taken prisoner.

'The Three Chins are the central Kingdoms. The temperament of the people is peaceable. The governments are orderly, and the peoples worn out by fighting. Although well trained for war they are contemptuous of their generals. They are stingy with officials' salaries. The officers have no will to fight to the death. Therefore, although well administered, they are of little use. The way to attack them is to press upon their formations and keep the pressure on. When their hosts arrive, resist them: when they withdraw, pursue them. So you can wear down their armies. Such is the general situation.

3. 'Naturally in an army there are certain to be some officers as brave as tigers and strong enough to lift a bronze tripod with ease; swift as wild horses they will capture battle flags and take generals prisoner. There are certain to be some who are able. These types should be selected, marked, cherished and honoured, for they are the life of the army. Such men are skilled in using the five weapons; talented, strong, nimble, and ambitious to swallow the enemy. You must increase their honours; they win decisive victories. Treat their parents, wives, and children liberally. Stimulate such officers with rewards and awe them with punishments, for they can endure in battle. If you are able to make a careful estimate of the abilities of such men they can attack twice their number.'

Marquis Wu said: 'Excellent.'

### SECTION II

1. Wu Ch'i said: 'Now in estimating the enemy situation there are eight conditions in which you may fight him without recourse to divination.

First, in times of strong wind and great cold when his men are awakened early to move and break ice to ford the rivers and do not shrink from hardship.[1] Second, in the scorching heat of mid-summer when they rise late and are pressed for time, and when it is necessary to march a great distance and they suffer from hunger and thirst. Third, when his army has already been encamped for a long time and is out of grain and food, when his people are resentful and angry and there are many evil omens and portents and the superior officers cannot put a stop to spreading rumours. Fourth, when the army's equipment is worn out and he is short of firewood and fodder, when there is constant mist and rain, and the troops wish to plunder but there is no opportunity. Fifth, when the army is not large, where the terrain and the water supply is inconvenient, where both men and horses are ill, and where the neighbouring states do not come to their aid. Sixth, when the roads are distant, the sun setting, and officers and men tired and apprehensive. They are exhausted and have not yet eaten nor have they taken off their armour to rest. Seventh, when the general is lax and his deputies negligent, when the officers and men are not firm in purpose and the host constantly alarmed; when the army is isolated and without aid. Eighth, when the enemy array is not drawn up, when he has not completed his encampment, or when he is marching in hilly country, or ascending precipices, when half are hidden and half seen. In all these conditions you can attack an enemy without hesitation.

2. 'There are six situations in which, without divining, you must avoid attacking the enemy.

---

[1] The text reads so, but this does not seem consistent.

These are, first, where his country is extensive and his people many and prosperous. Second, where the superiors love their inferiors and their benevolence grows and spreads. Third, where rewards are reliable and punishment is carefully considered and where these are administered appropriately. Fourth, where those who display merit are given suitable positions, where responsibilities are given to the wise and employment to the able. Fifth, where the army is large and well equipped. Sixth, when there is aid from all sides and the enemy is assisted by powerful states. Generally when unequal to your enemy in these matters, you must without doubt avoid him. What I mean is that when there is an opportunity, you may advance; when you see things are difficult, retire.'

### SECTION III

1. Marquis Wu inquired: 'I wish to observe the enemy's external appearance and so know his internal situation; to examine his advance and know when he will stop and thus determine the outcome. Can I hear how to do so?'

Wu Ch'i replied: 'When the enemy approaches carelessly and without a plan, when his flags and banners are confused and disorderly, when both men and horses often look to the rear, one can attack an enemy force ten times his own and surely rout it.

'When the forces of the feudal lords have not yet assembled, when sovereigns and ministers are not in accord, when moats and ramparts are not yet completed, when prohibitions and commands are not yet published, when the entire host is in an uproar, when they wish to advance and cannot, or to retire and do not dare, then one may attack an enemy twice his size, and in

one hundred battles there will be no calamity.'

2. Marquis Wu inquired: 'Under what conditions may an enemy certainly be attacked?'

Wu Ch'i replied: 'Employment of troops must be in accord with determination of the enemy's strong and weak points, after which you speedily attack his critical positions. When the enemy approaches from afar and has not yet arrayed his ranks and files, he can be attacked. When his troops have just eaten and are not prepared you can attack; when he is in flight you can attack; when he is exhausted and fatigued you can attack. When he has not yet secured advantage of the ground you can attack. When he has let the moment slip by or does not take advantage of opportunity to follow up, you may attack. When he has come a long way and is not rested you may attack. When in a river passage with but half his force across, you may attack; where the road is precipitous and constricted you may attack. When his flags and banners move confusedly you can attack; when his formations move about incessantly you may attack. When the generals become separated from the troops you can attack; when they are frightened you may attack. Now, in all such conditions you must employ *élite* troops to burst into the enemy's ranks; then divide your troops to follow up. Attack with speed and without hesitation.'

## CHAPTER III
### THE CONTROL OF TROOPS

### SECTION I

1. Marquis Wu asked: 'What is of first importance in the employment of troops?'

Wu Ch'i replied: 'First there should be clear understanding of the four qualities described as

"Easy", the two described as "Heavy", and the one described as "Reliability".'

2. The Marquis said: 'How do you describe these? What do you mean by them?'

Wu Ch'i replied: 'The ground should be easy for the horses; the horses should pull the chariots with ease; the chariots should carry the men with ease; the men should enter battle with ease. If you know the difficult and easy ground, ground is easy for the horses. If fodder and grain are provided at the right times the horses will draw the chariots easily. If there is an abundance of axle grease the chariots can easily carry the men; if their weapons are keen and their armour strong the men will fight with ease.

'There is heavy reward for an advance and heavy punishment for retreat.

'In the administration of rewards and punishments there must be reliability.

'One able to investigate these matters and fully apprehend them is a master of victory.'

3. Marquis Wu asked: 'What makes troops gain the victory?'

Wu Ch'i said: 'It is proper discipline that enables them to win victories.'

The Marquis then asked: 'Does it not lie in numerical strength?'

Wu Ch'i replied: 'If laws and orders are not clear and rewards and punishments not reliable, troops will not stop at the sound of the bells nor advance at the roll of the drums and though there be a million of this sort, of what use are they? What is called discipline is that when encamped their conduct is proper; that when on the move the army is awe-inspiring, so that in advance it cannot be opposed, in retirement it cannot be pursued. In advance or retirement it is in good order, both right and left wings respond to the signals given by banners. Though cut off they can reform; though dispersed they retain their files. Whether the position is secure or perilous the troops can be assembled and cannot be isolated. They can be used and not wearied. They can be thrown in any direction and none under Heaven can oppose them. An army of this sort is called a "Father–Son Army".'

### SECTION II

1. Wu Ch'i said: 'Now the method of conducting the march of an army is this. The established order of advancing and of halting is not to be violated; the right times for eating and drinking should not be missed. Do not deplete the strength of men and animals. In these three matters the troops must have confidence in the orders of their seniors. The orders of their superiors is the source whence discipline is born. If advances and halts are not well regulated, if food and drink are not suitable, the horses will be exhausted and the men fatigued and not able to be relaxed and sheltered. Such are the situations in which troops do not trust the orders of their superiors. If orders of superiors are ineffective, when the army encamps it will be disorderly, and when it fights, defeated.'

2. Wu Ch'i said: 'Now the field of battle is a land of standing corpses; those determined to die will live; those who hope to escape with their lives will die.

'A general good at commanding troops is like one sitting in a leaking boat or lying under a burning roof. For there is no time for the wise to offer counsel nor the brave to be angry. All must come to grips with the enemy. And therefore it is said that of all the dangers in employing troops, timidity is the greatest and that the calamities which overtake an army arise from hesitation.'

## SECTION III

1. Wu Ch'i said: 'Now men generally die when they cannot help it and are defeated by a disadvantageous situation. Therefore in employing troops, instructing them and warning them is of the first importance.

If one man studies war he can successfully instruct ten; if ten study they can successfully instruct a hundred; one hundred can successfully instruct one thousand, and one thousand can successfully instruct ten thousand. Ten thousand can instruct the entire army.

'Close to the field of battle, await an enemy coming from a distance; with fresh troops await an exhausted enemy, and with well fed troops a hungry one.

'In training, cause your troops to form squares from circles, to sit down and get up, to move and to stop. Cause them to pass from the left to the right and from the front to the rear. Divide and concentrate them, unite and disperse them. When versed in all these varying circumstances give them their weapons. These matters are described as the business of the general.'

2. Wu Ch'i said: 'The regulations for combat training are that the short men carry lances and halberds, and the tall men bows and crossbows. The strong carry the banners and flags; the valiant the bells and drums; the weak are servants and prepare food. The wise lay plans.

'Put men from the same villages together and the sections of ten and the squads of five will mutually protect one another.

'At one beat of the drum all put their weapons in order. At two beats they practise formations. At three beats they go speedily to meals. At four they prepare for action. At five they march off. When they have heard the drums

beat and listened to the orders, the banners are unfurled.'

3. Marquis Wu asked: 'Are there methods for controlling the army in the advance and at halts?'

Wu Ch'i replied: 'Do not face a Heavenly Oven or a Dragon's Head. A Heavenly Oven is the mouth of a great valley; a Dragon's Head, the end of a great mountain.

'Over the left wing flies the banner of the Green Dragon; over the right, that of the White Tiger; over the van that of the Red Bird, and over the Rear the banner of the Black Tortoise. Over the commander flies the banner of the Great Bear. Under it his staff is assembled. When about to fight, carefully investigate the direction of the wind. March with a following wind. When the wind strongly opposes you, await a change.'

4. Marquis Wu asked: 'Now in taking care of horses what is the method?'

Wu Ch'i replied: 'Now horses must have peaceful places to rest. Their drink and fodder must be suitable and they must be regularly and properly fed. In winter their stables must be warm, and in summer, cool and shady. The hair of their manes and tails must be cut and their four hooves carefully trimmed. Their eyes and ears should be controlled; do not permit them to stampede. They must be practised in galloping and pursuing. Their movement and stops should be leisurely. When mutual affection exists between the rider and his horse they can then be used.

'The equipment of the chariot horse and the cavalry mount consist of saddle, bridle, bit, and reins. These must be strong.

"Now in general, horses are not harmed toward the end of the march but from the beginning. They are not harmed by feeding them too little but are assuredly harmed when overfed. Toward sunset after a long march the riders must dismount and remount many times, for it is better to tire the men. Take heed not to tire the horses! Always see that they have a reserve of strength, and so prepare against the enemy's sneak attacks. One able to understand these matters will be unopposed in the world.'

## CHAPTER IV
### A DISCUSSION OF GENERALSHIP

#### SECTION I

1. Wu Ch'i said: 'Now the commander of an army is one in whom civil and martial acumen are combined. To unite resolution with resilience is the business of war.

'Usually when people discuss generals they consider only courage. Courage is but one of many qualities of generalship. Now a courageous man is certain to engage recklessly and without knowing the advantages. This will not do.

'Now there are five matters to which a general must pay strict heed. The first of these is administration; the second, preparedness; the third, determination; the fourth, prudence; and the fifth, economy. Administration means to control many as he controls few. Preparedness means that when he marches forth from the gates he acts as if he perceives the enemy. Resolution means that when he approaches the enemy he does not worry about life. Prudence means that although he has conquered, he acts as if he were just beginning to fight. Economy means being sparing in laws and orders so that they are not vexatious.

'To receive his orders and not to decline; to talk of return only after the enemy has been defeated, is the proper conduct of a commander-in-chief. Therefore on the day the army sets forth he has in mind the glory of death and not the shame of living.'

2. Wu Ch'i said: 'Now in employing troops there are four potentialities. The first is in respect to morale; the second in respect to terrain; the third in respect to situation; and the fourth in respect to strength. Now as for the multitude of the three armies, a host of a million, what is important is the responsibility of one man. This is called potential in respect to morale. Where roads are constricted and dangerous, where there are famous mountains and great bottlenecks and where if ten men defend, a thousand cannot pass, this is potential in respect to terrain.

Be skilful in employing spies and agents. Have the light troops manoeuvre in order to divide the enemy host. Make his sovereign and ministers mutually resentful, and his superiors and subordinates blame one another. This is called potential in respect to situation. When the chariots have strong axles and linch pins, the boats efficient oars and rudders, the troops training in combat formations and the horses training in swift pursuit, this is called potential in respect to strength. One who thoroughly comprehends these four matters can be made a general. Still, his majesty, virtue, humanity and courage must be sufficient to lead those under him and to give peace to the multitudes. He awes the enemy and causes him to be perplexed. When he issues orders none dares disobey, and wherever he is no rebels dare oppose him. If one gets a general like this his country is strong; if he dismisses one like this his country perishes. This is what is called The Excellent General.'

## SECTION II

1. Wu Ch'i said: 'Now horse-borne drums, ordinary drums, bells, and bells with clappers are used to impress the ears, and flags, banners, pennants, and streamers to impress the eyes. Prohibitory orders and punishment of crime impress the heart. When the ears are impressed by sounds they cannot but understand; when the eyes are impressed by colours they cannot but distinguish them; when the heart is impressed by punishment it cannot but be sternly controlled.

'Now when these three matters are not well established, any country is certain to be defeated by its enemies. And therefore it is said that where the general's banners are, there are none who do not follow, where the general points there are none who will not advance in the face of death.'

2. Wu Ch'i said: 'Generally a most important matter in war is to inquire concerning the enemy's generals and exhaustively examine their ability to act in accordance with circumstance. Then without tiring yourself your affairs may proceed.

'If the enemy general is stupid and places his confidence in others you can deceive him and lure him into traps. If he is covetous and careless of his reputation, you can bribe him.

'If he is easily changeable and lacks plans, you can tire him out and wear him down.

'If the superiors are rich and arrogant and the inferiors poor and resentful, you can divide and separate them.

'If the general is hesitant in advancing or retiring, his host will have no confidence in him, and can be stampeded and put to flight. If the officers are contemptuous of him and are of a mind to return home, block the easy routes, and open to them those that are difficult. Then you can intercept and capture them.

'When the road by which he advances is easy and that by which he would retire difficult you can bring him to your front. When the road by which he advances is dangerous and that by which he would retire easy, you can approach and attack him.

'If he encamps in low-lying damp ground where the drainage is poor and rain falls in abundance, you can inundate and drown him. If he encamps in a desolate marsh or in a place where violent winds frequently blow you can set fires and burn him out.

'If he remains, long in one place without moving, the generals and officers become indolent and remiss and his army will not be prepared, and you can approach secretly and strike him.'

## SECTION III

1. Marquis Wu asked: 'When the opposing armies confront one another and we know nothing about the enemy general, and I wish to determine his qualities, what devices are suitable?'

Wu Ch'i replied: 'Order bravos in command of some *élite* troops to try him out. Their sole purpose is to flee, not to gain anything but to observe how the enemy reacts. If his actions are in unison and his discipline good, and when he pursues and pretends to be unable to catch up, when he sees an advantage but pretends to be unaware of it, then the general is wise and you should not engage him.

'But if the enemy host is clamorous and bawling, its flags and banners confused and disorderly, the troops running and stopping without orders, their weapons held sometimes one way, sometimes another; when in pursuit of the flee-ing they are unable to catch up, when they see an advantage but are unable to take it, then the gen-eral is stupid and you can capture him.'

## CHAPTER V
### ON RESPONDING TO CHANGING CONDITIONS

#### SECTION I

1. Marquis Wu asked: 'When chariots are well built and horses fast, generals valiant and troops strong, and when you unexpectedly encounter the enemy, your forces fall into disorder and ranks are broken, what is to be done?'

Wu Ch'i replied: 'Now the system by which battle formations are regulated is that by day banners, flags, streamers, and pennants are used, and by night bells, drums, reed pipes, and flutes. When the pennants indicate "Left" the troops move to the left; when they indicate "Right" the troops move to the right. When the drums roll they advance; when the bells sound they stop. At one piping they form ranks; at the second they assemble. Those who do not obey the orders are punished.

2. 'If the entire army is thus subject to authority and officers and men put forth their utmost, then in battle there is no enemy strong enough to resist you and when you attack, no formation firm enough to withstand you.'

#### SECTION II

1. Marquis Wu asked: 'If a greatly superior enemy force attacks my inferior force, what rem-edy is there?'

Wu Ch'i replied: 'If the ground is easy, avoid him; in a defile, encounter him. For it is said that when one attacks ten, no place is better than a defile; for ten to attack a hundred nothing is bet-ter than a precipitous place. When one thousand attack ten thousand, nothing is better than a mountain pass. Now if you have a small force and suddenly attack the enemy in a narrow road with gongs sounding and drums rolling, his host, however large, will be alarmed.

'And therefore it is said that one employing large numbers seeks easy ground; one employing small numbers, constricted ground.'

#### SECTION III

1. Marquis Wu asked: 'Suppose there is a large army of high morale and, as well, valiant. To its rear are constricting passes, to its right moun-tains, to its left a river. It is well protected by deep moats and high ramparts defended by strong crossbowmen. In retirement it resembles a moving mountain; in advance, a blizzard. It is difficult to hold up such an army. What is to be done in these circumstances?'

Wu Ch'i replied: 'A good question! In this sit-uation the outcome does not depend on the strength of chariots or cavalry, but on the plans of a sage. You should make ready one thousand chariots and ten thousand horsemen together with infantry and divide them into five columns. Assign to each column a separate route. Thus, with your five columns on five different roads, the enemy is certain to be perplexed and will not know which to deal with.

'If the enemy has taken up a strong defensive position in order to consolidate his troops, quickly send *agents provocateurs* and spies to ascertain his plans. If he listens to their propos-als he may disengage and retire. If he does not he will execute the envoys and burn our letters.

'Then our five armies simultaneously engage. If victorious, we do not pursue; if not victorious, we retire rapidly simulating flight. We march steadily but ready to fight at any time. One column ties up his front while another cuts off his rear. Then two columns swiftly and silently suddenly attack a vulnerable spot. Sometimes to his left, sometimes to his right. If the five columns co-ordinate their attacks, it will assuredly be advantageous. This is the way to attack a strong enemy.'

### SECTION IV

1. Marquis Wu inquired: 'If the enemy is close by and presses upon me and I wish to retire but there is no road and my host is alarmed, what is the remedy?'

Wu Ch'i replied: 'The tactics to use in these circumstances are as follows: If we are many and the enemy few we divide our forces and charge. If the enemy is many and we few a method must be found to harass him. By harassing him unceasingly his host, however large, can be overcome.'

### SECTION V

1. Marquis Wu inquired: 'If we unexpectedly encounter the enemy in a gorge or valley with precipitous sides and he is vastly superior in numbers, what is to be done?'

Wu Ch'i replied: 'In all cases you must march rapidly to get away from such places as mountains, forests, valleys, and marshlands. You cannot march leisurely. When in high mountains or deep valleys you unexpectedly encounter the enemy, you must roll the drums and shout and seize the opportunity to shoot with bows and crossbows, thus wounding some and taking others prisoner. Carefully observe the state of his array and if he falls into confusion, strike without hesitation.'

### SECTION VI

1. Marquis Wu inquired: 'When to both right and left there are high mountains with constricted ground between and I unexpectedly encounter the enemy and dare not attack him and wish to withdraw but cannot, what may be done?'

Wu Ch'i replied: 'This is described as "valley fighting". Here, superiority in numbers avails nothing. You assemble the most talented among the officers to face the enemy and advance with well armed light troops in the van. Divide the chariots and cavalry into groups and conceal them on all sides, separated by several *li,* so that the enemy cannot see them. The enemy will surely entrench and will not venture either to advance or retire. Whereupon raise your banners and pennons, march out of the mountainous country and encamp. The enemy will certainly be apprehensive. Provoke him with your chariots and allow him no rest. Such is the method of fighting in valleys.'

### SECTION VII

1. Marquis Wu asked: 'Suppose I suddenly encounter the enemy in a flooded marsh. The chariot wheels sink in the muck and the shafts are submerged and water overwhelms both vehicles and horsemen. We are not equipped with boats and oars, and can neither advance nor retire. What then is to be done?'

Wu Ch'i replied: 'This is called "water fighting". Chariots and cavalry are of no use. You must keep them at one side. Ascend to a high

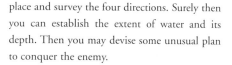

place and survey the four directions. Surely then you can establish the extent of water and its depth. Then you may devise some unusual plan to conquer the enemy.

'When the enemy is crossing a river, wait until half his force is across and then attack him.'

### SECTION VIII

1. Marquis Wu asked: 'When rain is unceasing, horses sink in the mud, the vehicles are bogged down, and the enemy attacks on all sides and my troops are alarmed and fearful, what may be done?'

Wu Ch'i replied: 'Generally, chariots are not used when the weather is cloudy and damp, but may be sent into action when it is bright and dry. High ground is to be preferred; low ground to be shunned. Hasten on with your best chariots. When advancing or halted you must follow the roads taken by the chariots. If the enemy moves you must follow him up.'

### SECTION IX

1. Marquis Wu inquired: 'Supposing fierce rebels arrive to plunder the countryside and carry off the domestic animals, what can be done?'

Wu Ch'i replied: 'When fierce bandits come you must take precautions against them. It is best to stand on the defensive and not meet them. When at dusk they are about to depart laden and encumbered with loot they are certain to be apprehensive, for their withdrawal is by distant routes and they must travel fast. There are certain to be some who lag behind. Follow them and strike and you can rout them.'

### SECTION X

1. Wu Ch'i said: 'Generally the method of attacking and besieging enemy cities is this. When his walled cities and towns have been broken into, enter all his encampments and take over his wealth and offices, his tools and animals.

'Where the army encamps you must not cut down trees, destroy dwellings, take away crops, slaughter the domestic animals, or burn the granaries.

'Thus you demonstrate to the people that you have no desire to oppress them. Those who wish to surrender should be allowed to do so, and permitted to live in peace.'

## CHAPTER VI
### ENCOURAGEMENT OF OFFICERS

### SECTION I

1. Marquis Wu asked: 'Are severe punishments and enlightened rewards sufficient to ensure victory?'

Wu Ch'i replied: 'As to the matter of severity in punishment and enlightenment in rewards, I am not able to comprehend the whole of it. However, they cannot be exclusively relied upon. Now to publish orders and make known commands which the troops happily obey, to raise the army and mobilize the people in such a way that men are happy to fight, and when they cross swords, happy to face death, are the three matters in which a sovereign places his reliance.'

2. Marquis Wu asked: 'How is this to be achieved?'

Wu Ch'i replied: 'When the sovereign raises up those who are meritorious and entertains them and when he encourages those who lack merit.'

3. Whereupon Marquis Wu set places for a banquet in the halls of the palace, and made provision for seating his guests in three ranks, and entertained the nobles and great officers at a banquet. Those most meritorious sat in the front row. Delicate meats were set before them, and beef in the finest vessels was offered them.

Those next in order of merit sat in the middle row. They were served delicacies in dishes less fine.

Those with no merit sat in the rear and were served delicacies in small vessels.

When the banquet was over and the guests had left, gifts were bestowed on the fathers, mothers, wives and children of the meritorious outside the palace gates, and the same discrimination was applied as previously. Envoys were sent each year to thank and bestow gifts upon the fathers and mothers of those killed in war to testify that they were held in remembrance.

When this policy had been in effect for three years, the Ch'in people set forth with their army and approached the west bank of the river. When the officers of Wei heard this, those who did not await orders from the officials but donned their armour and attacked the Ch'in vigorously numbered several tens of thousands.

4. The Marquis of Wu summoned Wu Ch'i and said to him: 'We see now the effect of your teachings.'

Wu Ch'i replied: 'I have heard that some men are more talented, others less so; that spirits are at times flourishing and at others depressed. Now if you experiment by sending forth fifty thousand men of no merit, I beg to lead them to oppose the enemy. Suppose I do not win, I will be the laughing stock of the feudal lords, and we will lose all our authority in the Empire. Now suppose there is a desperate bandit lurking in the fields and one thousand men set out in pursuit of him. The reason all look for him as they would a wolf is that each one fears that he will arise and harm him. This is the reason one man willing to throw away his life is enough to terrorize a thousand.

'Now if I use a host of fifty thousand and all are like this one desperate bandit and I lead them to attack the Ch'in, certainly they cannot match me.'

5. Whereupon Marquis Wu followed this advice, and with five hundred chariots together with three thousand cavalry defeated a Ch'in army of half a million. This illustrates the efficacy of encouraging officers.

6. On one day prior to battle Wu Ch'i gave orders to the armies in the following terms: 'All officials and officers should follow up and capture the enemy's chariots, cavalry and footmen. If the chariots do not capture chariots, the horsemen capture cavalry and the footmen capture infantry, even if the enemy army is destroyed no one will acquire any merit.' Therefore on the day of battle his orders were not vexatiously numerous, but his prestige shook the Empire.

# BRIEF BIOGRAPHIES
# OF THE
# COMMENTATORS

**Ts'ao Ts'ao** (AD 155–220) was made King of Wei by the Han Emperor Hsien Ti in AD 216. He died in AD 220 and was canonized 'Martial King' in AD 237. After his son became Emperor, this title was changed to 'Martial Emperor' with the Temple designation T'ai Tzu, or 'Eminent Founder' (of the Wei Dynasty).

His biography as given in The Wei Dynastic History (Wei *Shu*) is quoted in *The Chronicle of the Three Kingdoms*:

T'ai Tsu, since he governed the whole empire, mowed down numerous scoundrels. In his military operations, he followed in the main the tactics laid down in the Sun-tzu and Wu-tzu. In accordance with different situations, he took extraordinary stratagems; by deceiving the enemy, he won victory; he varied his tactics in demonic fashion. He himself wrote a book on war, consisting of a hundred thousand and several tens of thousands of characters, and when his generals undertook any campaign they all followed this new book. Furthermore, on each occasion he gave them personal directions; those who obeyed them won victory, and those who did not were defeated. In the face of the enemy on the battlefield, he remained unperturbed, as if he had no intention whatever of fighting; but seizing his opportunity, he would strike for victory in the highest spirits. This is why he always won victory whenever he fought, not a single instance of his successes being attributed to mere good luck. He knew men well and was adept in judging them; it was difficult to dazzle him by false display. He picked Yü Chin and Yüeh Chin out from the rank and file, and Chang Liao and Hsü Huang from among the surrendered forces; all of them became his supporters and achieved merit, becoming famous generals. Furthermore, the number of those whom he picked up from mean and insignificant positions, and who eventually rose to be governors of provinces and prefects, cannot be counted. It was thus that he laid the foundations of his great work. He cultivated both the art of peace and the art of war: during the thirty-odd years when he commanded troops, books never left his hand. During the day he attended to military matters, during the night he applied his mind to the Classics and their commentaries. When he climbed a height, he would always compose verses. When he made new poems, he would set them to pipe and string, and they all turned out to be excellent songs. His talents and strength were unsurpassed; with his own hands he could shoot down flying birds and capture ferocious beasts alive. Once he shot down sixty-three pheasants in a single day at Nan-p'i. When palaces were constructed and machines repaired, he always laid down rules which proved to work to the utmost satisfaction. By nature he was temperate and frugal, not given to pomp and adornment. Ladies of his harem did not wear any embroidered garments, his attendants did not have two pairs of footgear. When his coloured curtains and wind-screens were damaged, he had them patched; he had his bedding only for keeping warm, devoid of border ornament. All things of beauty and elegance which he obtained as booty from captured cities and towns, he would distribute among those who had shown merit. In acknowledging and rewarding services, he was not one to consider a thousand gold pieces too much; but to those without merit who sought to profit from his largesse, he would not give a single cash. Gifts presented to him from the four quarters, he shared with his subordinates. He was of the opinion that the funeral service of the time was too extravagant and useless, the vulgar carrying it to excess; he therefore made a stipulation as to his own funeral, that no more than four basketsful of clothing were to be buried with him.

Elsewhere he is not treated in the same favourable light:

. . . But in the maintenance of laws he was harsh and exacting. If any of his subordinate generals had better counsels of war than his, he would find an opportunity to put him to death under the pretext of some law; and none of his former associates and friends who had earned his grudge were spared alive. When he put a man to death, he used to look at him, weeping and lamenting over him, but he would never grant a pardon.

*(The Chronicle of the Three Kingdoms,* vol. i, pp. 1, 15, 16, 17.)

**Tu Yü** (AD 735–812) A native of Wan-nien in Shensi, who rose to be President of the Board of Works and Grand Guardian. He compiled the *T'ung Tien,* an encyclopedic work divided into eight sections; Political Economy, Examinations and Degrees, Government Offices, Rites, Music, Military Discipline, Geography, and National Defence. He was ennobled as Duke of Ch'i-Kuo and canonized after his death.

**Li Ch'üan**  A T'ang writer on military subjects. His principal works were *T'ai Pai Yin Ching* and *Chiang Lüeh,* both of which are extant.

**Tu Mu**  (AD 803–52) A native of Wan-nien, Tu Mu graduated as *chin shih* about 830. He rose to be a secretary in the Grand Council. As a poet he achieved considerable distinction and is often spoken of as the Younger Tu, to distinguish him from Tu Fu. His biography is attached to that of his grandfather in the New T'ang Dynastic History, chapter 166.

**Mei Yao-ch'en**  (AD 1002–60) A native of Wan-Ling in Anhui, and a famous Sung poet. In 1056 he was summoned to the Imperial Academy and rose to be a second-class secretary. In consequence of his work on the T'ang dynasty, he was placed on the commission to prepare the New History of that period, but died before its completion. In addition to his commentary on Sun Tzu he wrote several explanatory works on The Book of Odes. After his death the noted literary critic Ouyang Hsiu composed an Epitaph or Eulogy. His biography appears in the Sung Dynastic History, chapter 443.

**Wang Hsi**  A native of T'aiyuan in present Shansi province. He was a Hanlin scholar and a government official. His principal literary interest was The Spring and Autumn Annals on which he wrote several critical essays.

**Chang Yü**  A late Sung historian and critic. His principal work was *Pai Chiang Chuan: The Biographies of One Hundred Generals.* Nothing more is known of him.

Nothing is known of the careers of the early commentators identified only as 'Mr.' Mêng, or of Ch'ên Hao, Chia Lin, and Ho Yen-hsi.

# BIBLIOGRAPHY

## I

### BOOKS IN ENGLISH

Aston, W.G. *The Nihongi.* Transactions and Proceedings of the Japan Society, Supplement I. London, 1896. Kegan Paul.

de Bary, William Theodore, and others. *Sources of Chinese Tradition.* New York, 1960. Columbia University Press.

Baynes, Cary F. *The I Ching,* or *Book of Changes.* The Richard Wilhelm Translation. London, 1951. Routledge & Kegan Paul Ltd.

Calthrop, Captain E.F. *The Book of War.* London, 1908. John Murray.

Carlson, Evans F. *Twin Stars of China.* New York, 1940. Dodd, Mead & Co.

Cheng, Lin. *The Art of War.* Shanghai, China, 1946. The World Book Company Ltd.

Dubs, Professor Homer H. (trans.) *History of the Former Han Dynasty.* (3 vols.) Baltimore, Maryland, 1946, 1955. The Waverly Press.

— *Hsün Tze, The Moulder of Ancient Confucianism.* London, 1927. Arthur Probsthain.

— *The Works of Hsün Tze.* London, 1928. Arthur Probsthain.

Duyvendak, J.J.L. *Tao Te Ching. The Book of the Way and Its Virtue.* London, 1954. John Murray.

— *The Book of Lord Shang.* London, 1928. Arthur Probsthain.

Fitzgerald, C.P. *China, A Short Cultural History.* (rev. ed.) London, 1950. The Cresset Press Ltd.

Fung, Yu-lan. *A History of Chinese Philosophy.* (trans. Bodde) Princeton, 1952. Princeton University Press.

Gale, Esson M. (trans.) *Discourses on Salt and Iron.* Sinica Leidensia, vol. ii. Leiden, 1931. E. J. Brill Ltd.

Giles, Lionel. (trans.) *Sun Tzu on the Art of War.* London, 1910. Luzac & Co.

Granet, Marcel. *Chinese Civilization.* London, 1957. Routledge & Kegan Paul Ltd.

Legge, James. *The Chinese Classics.* London, 1861. Trubner & Co.

Liang, Ch'i-ch'ao. *Chinese Political Thought.* London, 1930. Kegan Paul; Trench, Trubner & Co. Ltd.

Liao, W.K. (trans.) *The Complete Works of Han Fei-tzu* (2 vols.). London, 1939 (vol. i); 1959 (vol. ii). Arthur Probsthain.

McCullogh, Helen Craig. (trans.) *The Taiheiki. A Chronicle of Medieval Japan.* New York, 1959. Columbia University Press.

Machell Cox, E. *Principles of War by Sun Tzu.* Colombo, Ceylon. A Royal Air Force Welfare Publication.

Mao Tse-tung. *Selected Works.* London, 1955. Lawrence & Wishart.

— *Strategic Problems in the Anti-Japanese Guerrilla War.* Peking, 1954. Foreign Language Press.

Mei, Y.P. *Motse, the Neglected Rival of Confucius.* London, 1934 Arthur Probsthain.

— *The Ethical and Political Works of Motse.* London, 1929. Arthur Probsthain.

Müller, Max F. (ed.) *The Sacred Books of the East* (vol. xv): *The Yi King.* (trans. Legge) Oxford, 1882. The Clarendon Press.

Murdoch, James. *A History of Japan.* (3rd impression) London, 1949. Routledge & Kegan Paul Ltd.

Payne, Robert. *Mao Tse-tung, Ruler of Red China.* London, 1951. Secker & Warburg.

Sadler, Professor A.L. *The Makers of Modern Japan.* London, 1937. George Allen & Unwin Ltd.

— *Three Military Classics of China.* Sydney, Australia, 1944. Australasian Medical Publishing Co. Ltd.

Sansom, George B. *A History of Japan to 1334.* (San II) London, 1958. The Cresset Press.

— *Japan, A Short Cultural History.* (2nd impression, revised) (San I) London, 1952. The Cresset Press Ltd.

Schwartz, Benjamin I. *Chinese Communism and The Rise of Mao.* (3rd printing) Cambridge, Massachusetts, 1958. Harvard University Press.

Snow, Edgar. *Red Star over China.* (Left Book Club Edition.) London, 1937. Victor Gollancz Ltd.

Tjan Tjoe Som (Tseng, Chu-sen). *The Comprehensive Discussions in The White Tiger Hall.* Leiden, 1952. E. J. Brill.

Tsunoda, Ryusaka; de Bary, William Theodore; and Keene, Donald. *Sources of Japanese Tradition.* New York, 1958. Columbia University Press.

Waley, Arthur. *The Analects of Confucius.* London, 1938. George Allen & Unwin Ltd.

Walker, Richard L. *The Multi-State System of Ancient China.* Hamden, Connecticut, 1953. The Shoe String Press.

Watson, Burton. *Ssu-ma Ch'ien, Grand Historian of China.* New York, 1958. Columbia University Press.

## II

## MONOGRAPHS AND ARTICLES IN ENGLISH

Bodde, Dirk. *Statesman, Patriot and General in Ancient China.* New Haven, Connecticut, 1943. A Publication of the American Oriental Society.

Chang, Ch'i-yün. *China's Ancient Military Geography.* Chinese Culture, vii, no. 3. Taipei, December 1959.

Extracts from China Mainland Magazines. 'Fragmentary Notes on the Way Comrade Mao Tse-tung Pursued his Studies in his Early Days.' American Consulate General. Hong Kong, 191, 7 December 1959.

Lanciotti, Lionello. *Sword Casting and Related Legends in China, I, II.* East and West, Year VI, N. 2, N. 4. Rome, 1955, 1956.

Needham, J. *The Development of Iron and Steel Technology in China.* London, 1958. The Newcomen Society.

North, Robert C. 'The Rise of Mao Tse-tung.' *The Far Eastern Quarterly,* vol. xi, no. 2, February 1952.

Rowley, Harold H. *The Chinese Philosopher Mo Ti* (reprint from *Bulletin of the John Rylands Library,* vol. xxxi, no. 2, November 1948). Manchester, 1948. The Manchester University Press.

Selections from China Mainland Magazines. 'Comrade Lin Piao in the Period of Liberation War in the Northeast.' American Consulate General, Hong Kong, 217, 11 July 1960.

Teng, Ssu-yü. *New Light on the History of the T'aip'ing Rebellion.* Cambridge, Massachusetts, 1950. Harvard University Press.

Van Straelen, H. *Yoshida Shoin.* Monographies du T'oung Pao, vol. ii. Leiden, 1952. E. J. Brill.

# III

# BOOKS, MONOGRAPHS AND ARTICLES IN WESTERN LANGUAGES (OTHER THAN ENGLISH)

Amiot, J.J.L. *Mémoires concernant l'histoire, les sciences, les arts, les moeurs, les usages, etc. des Chinois.* Chez Nyon l'aîné. Paris, 1782.

Ashiya, Mizuyo. *Der Chinesische Kriegsphilosoph der Vorchristlichen Zeit.* Wissen und Wehr, 1939, pp. 416—27.

Chavannes, Edouard. *Les Mémoires historiques de Se-ma Ts'ien.* Paris. Ernest Leroux.

Cholet, E. *L'Art militaire dans l'antiquité chinoise.* Paris, 1922. Charles-Lavauzelle.

Cotenson, G. de. 'L'Art militaire des Chinois, d'après leurs classiques.' *Le Nouvelle Revue.* Paris, August 1900.

Gaillois, Brig.-Gen. R. *Lois de la guerre en Chine.* Preuves, 1956

Konrad, N.I. *Wu Tzu.* Traktat o Voennom Isskusstve. Moscow, 1958. Publishing House of Eastern Literature.

— *Sun Tzu.* Traktat o Voennom Iskusstve. Moscow, 1950. Publishing House of the Academy of Science USSR.

Maspero, Henri. *La Chine Antique.* (Nouvelle éd.) Paris, 1955. Imprimerie Nationale.

Nachin, L. (ed.) *Sun Tse et les anciens Chinois Ou Tse et Se Ma Fa.* Paris, 1948. Editions Berger-Levrault.

Sidorenko, J.I. *Ssun-ds' Traktat über Die Kriegskunst.* Berlin, 1957. Ministerium Für Nationale Verteidigung.

# IV

# WORKS IN CHINESE

*Chan Kuo Shih.*
戰 國 史
'A History of the Warring States.' Yang K'uan. People's Press. Shanghai, 1956.

*Chao Chu Sun Tzu Shih San P'ien.*
趙 註 孫 子 十 三 篇
'Chao (Pen-hsueh's) Commentary on the Thirteen Chapters of Sun Tzu.' Chao Pen-hsüeh. Peiyang Military Academy Press. Peking, 1905.

*Chin I Hsin P'ien Sun Tzu Ping Fa.*
今 譯 新 編 孫 子 兵 法
'A Modern Translation of Sun Tzu's Art of War with New Chapter Arrangement.' Kuo Hua-jo. People's Press. Peking, 1957.

*Ch'in Ting Ku Chin T'u Shu Chi Ch'eng.*
欽 定 古 今 圖 書 集 成
Short Title: *T'u Shu.* 'Photographic reproduction of the Palace Edition of 1731. Chapter 83 'Military Canon'. Chung Hua Shu Chü. Shanghai, 1934.

*Chung Kuo Ping Ch'i Shih Kao.*
中 國 兵 器 史 稿
'A Draft History of Chinese Weapons.' Chou Wei. San Lien Shu Tien. Peking, 1957.

*Ku Chin Wei Shu K'ao Pu Cheng.*
古 今 僞 書 考 補 証
'A Further Inquiry into Apocryphal Books both Ancient and Modern.' Huang Yun-mei. Shantung People's Press, 1959.

*Pei T'ang Shu Ch'ao.*
北 堂 書 鈔
'Selected Passages Transcribed in the Northern Hall.' Yü Shih-nan (558—638).

*Shih Ch'i Hsüan.*
史記選
'Selections from the Historical Records.'
Wang Po-hsiang. People's Literary Press.
Peking, 1958.

*Ssu-ma Fa.*
司馬法
'The Art of War of Ssu-ma Jang-chiu.'
*Ssu Pu Pei Yao* ed. Chung Hua Shu
Chü. Shanghai.

*Sun Tzu Shih San P'ien Chiao Chien
Chi Yao.*
孫子十三篇校箋舉要
'Notes on the Collation of Sun Tzu's
Thirteen Chapters.' Yang P'ing-an.
*Peking University Journal of the
Humanities,* no. i, 1958.

*Sun Tzu Chi Chiao.*
孫子集校
'A Collated Critical Study of Sun Tzu.'
Yang P'ing-an. Chung Hua Shu Chu.
Shanghai, 1959.

*Sun Tzu.*
孫子
The Sun Tzu with Commentaries.
Sun Hsing-yen. *Ssu Pu Pei Yao* ed.
Chung Hua Shu Chü. Shanghai, 1931.

*Sun-Wu Ping Fa.*
孫吳兵法
'The Arts of War of Sun (Tzu) and
Wu (Ch'i).' Ta Chung Shu Chü.
Shanghai, 1931.

*T'ai P'ing Yü Lan.*
太平御覽
Li Fang, 3rd series. *Ssu Pu Tsung K'an,*
chs. 270–359. Commercial Press.
Shanghai, 1935.

*T'ung Chih.*
通志
Cheng Ch'iao. Facsimile Reproduction
of Palace ed. of 1859, ch. 68.

*T'ung Tien.*
通典
Tu Yu. Facsimile Reproduction of
Palace ed. of 1859, chs. 148–62.

*Wei Shu T'ung K'ao.*
僞書通考
'A Comprehensive Study of Apocryphal
Books' (rev. ed.). Chang Hsin-cheng.
Commercial Press. Shanghai, 1957.

*Wu Ch'i Ping Fa.*
吳起兵法
'The Art of War of Wu Ch'i.' *Ssu Pu
Pei Yao* ed. Chung Hua Shu Chü.
Shanghai, 1931.

*Wu Ching Tsung Yao.*
武經總要
'Essentials of the Martial Classics.'
Tseng Kung-liang. *Ssu K'u Ch'uan
Shu* ed.

# INDEX

# PICTURE CREDITS

**Caption information:**

**Page 90:** A Chinese admiral, from an account of the sea voyages of Zheng He (1371–1433). **94:** A Tang-period painted stone depiction of two horsemen. **101:** The defensive tower and walls at Jiayuguan on the Silk Road. **104:** A soapstone representation of the god of war, Kuan Ti (Guan Di). **108:** Assorted goods being transported for the emperor, from the 17th-century *History of the Emperors of China* (colour on silk). **112:** A detail showing captives kneeling, from *The First Emperor of the Han Dynasty Entering Kuan Tung* by Chao Po-chu. **114:** An atlas manuscript showing the roads, military posts and strategic points in the fortifications of Canton and parts of Zhejiang and Jiangsu provinces. **119:** A Chinese general surrenders the mineral-rich city of Fushun in 1618, from a facsimile of the 1781 edition of the *Manzhou shilu* by Men Yingzhao. **126:** A review of troops, from an 18th-century handscroll, ink and colour with gold on silk. **130:** A bronze *kui* dagger-axe from Eastern Zhou, Warring States period, decorated with a relief *taotie* monster-face and a geometric pattern. **134:** A 13th-century album leaf depicting powerfully breaking ocean waves: ink on silk. **138:** Han-dynasty bronze mirror whose decorations reveal the interconnectedness of cosmic energy forces. **143:** A pair of pottery figures of protective soldiers, Northern Wei dynasty. **144:** Mounted warriors in combat, from the *San guo zhi zhuan* (*Romance of the Three Kingdoms*), set during the 2nd century AD. **149:** A Chinese soldier in full military regalia; 19th century. **154:** A parade of cavalry troops, from a handscroll. **159:** The entourage of the First Emperor travelling through the mountains. **162:** A set of Tang-dynasty *sancai* pottery equestrian musicians. **164–165:** A fortress along the Great Wall at sunset. **170:** Map of the walled city of Jiujiang and environs, 19th century. **174:** Tiled rooftops within the ancient city of Pingyao. **177:** A scroll illustration of a military leader – said by some to be the legendary Yellow Emperor (Huang Ti). **180:** Terracotta warriors from the tomb of the First Emperor at Xi'an. **185:** A waterfall on the Zhenzhu River in Sichuan. **188–189:** An 18th-century painting showing troops being sent into battle. **194:** *Mansions in the mountains of paradise*, a 10th-century scroll painting by Tung Yuan. **199:** Fertile terraced fields in western China. **202:** The haunting, mountainous landscape of Anhui. **206:** *The Journey of Emperor Ming Huang to Shu*, Northern landscape painting. **210:** A detail from *Saying Farwell at Hsun Yang* by Chiu Ying 1494–1552. **216:** A bronze horse-drawn carriage with driver, Eastern han dynasty, 2nd century AD. **221:** Soldiers and advisers form the retinue of a powerful leader. **224:** The powerful imperial symbol of the dragon, accompanied by elemental lightning flashes. **226–227:** An 11th-century handscroll showing the conflict between the divine forces of Huang Ti and Chi You, using fire and wind. **228:** Exploding Arrows, from *Wubeizhi* (*On Warfare*) by Mao (Yuanyi), a woodblock print book about all aspects of Chinese warfare, 18th–19th century. **230:** Men exchanging words on a remote path, from a carved celadon boulder. **234:** A scene from the *San guo zhi zhuan* (*Romance of the Three Kingdoms*), set during the 2nd century AD.